Ooey Gooey Chemistry

Curriculum for Homeschool and Co-op Students

By Marie Ecker

Easy Peasy Science Lab Curriculum

Printed in the United States of America

First Printing, 2017

Revised Printing, 2022

Edited by Mike Ecker
Cover Illustration: Pigknit

Easy Peasy Science
Satellite Beach, FL 32937

Table of Contents

Introduction to Science & Chemistry

SCIENCE SAFETY RULES

In science, we do a lot of great hands-on labs. It is important to follow these rules to keep yourself and others safe! Safety is always our priority in science class.

1. Read through any instructions or lab procedures before beginning an experiment. Listen carefully to verbal instructions.
2. If you do not understand how to do something, ask for help.
3. Do not play around or run in the lab when doing an experiment.
4. Immediately let your teacher know about unsafe conditions or behaviors you observe.
5. Do not touch equipment or chemicals until you are instructed to do so.
6. Always use science tools the way they should be used.
7. Do not eat or drink anything while completing a lab. Do not eat or drink anything out of your lab equipment.
8. Clean up spilled substances immediately.
9. Never use cracked or broken glass. Inform your teacher or parent if is broken.
10. Label any container you put chemicals in.
11. Never pour liquids into containers held in your hands.
12. Do not look directly into beaker or test tube. Hold up the container and look from the side.
13. Smell a substance by fanning the smell toward you with your hand. Do not put it directly to your nose.
14. Tie back long hair. Wear closed toed shoes. Do not wear loose clothing or jewelry.
15. Wear goggles, apron and gloves when appropriate.
16. Keep hands away from face, eyes, mouth and body while using chemicals or lab equipment.
17. If a chemical gets into your eyes or on your skin, rinse the area for 15 minutes with water.
18. Wash your hands carefully after each activity and after handling chemicals.
19. Keep your work area neat while working.
20. Clean up your area and equipment when you are finished. Put chemicals away correctly and in a safe place where young children cannot reach them.
21. Know what to do and be prepared in case of an emergency. Know the location of safety equipment and the exits. Follow emergency rules and protocol when accidents happen.

Hazard Symbols

Some chemicals can harm us if they are not handled carefully. Hazard symbols are used to identify the danger posed by each chemical and what care should be taken.

Irritant/Harmful: Skin or respiratory irritation.

Caution: Blistered skin or rashes.

Corrosive: Can attack and destroy living tissue, such as skin or bone.

Toxic: Chemicals which are poisonous and can cause death if they enter the body.

Flammable: Burn or ignite easily. Must be stored safely and used with care.

Explosive: Substances have enough stored energy in them that if treated incorrectly may explode.

High Voltage: Electrical energy at voltages high enough to inflict harm on living organisms.

Gas under pressure. May be cold when released, and explosive when heated.

Safety Contract

- I have read and understand the science safety rules. I agree to follow them throughout the course of this class.
- I will listen carefully to my teacher and follow all directions.
- I will wear the correct safety equipment when performing experiments.
- I agree to be responsible and to act with caution when performing experiments and activities.

______________________________ ____________
Student Signature Date

______________________________ ____________
Parent Signature Date

Oops Lab

Name:

Describe what is wrong with each numbered spot and the lab rule/rules that have been broken. However, one of the students is doing everything right. Which number are they?

1	**9**
2	**10**
3	**11**
4	**12**
5	**13**
6	**14**
7	**15**
8	**16**

Mentos and Coke Experiment

Name:

We are going to follow the scientific method as we test the chemistry between Mentos and Coke!

Goggles and Aprons on!
This can be a messy lab so dress appropriately. This is a good lab to do outside.
Liquid can shoot out of the bottles so it is important to protect your eyes.

Question: Will the reaction between Mentos and regular Coke create a taller geyser than the reaction with Mentos and Diet Coke?

Background Research:

- The original Mint Mentos are round candies with a chewy interior and a hard outer shell. The outside of mint Mentos are covered with many rough small pores because they are sprayed with over 40 microscopic layers of sugar. Gelatin and gum arabic are some of the ingredients in mint Mentos. Mentos are also heavy and will sink in liquid.
- To make your soda bubbly, carbon dioxide is forced into it at extremely high pressures. Diet sodas have slightly more carbonation (carbon dioxide) than regular sodas with different sweeteners.
- Mentos reacting with soda is a **physical reaction**, caused by the small pitted holes on the Mento. The Mento quickly attracts and releases a large amount of the carbon dioxide gas. Because it is heavy and sinks it does this throughout the bottle as it falls. The sudden increase in pressure from the gas being released is what pushes the liquid up and out of the bottle.
- A **Physical Reaction** or **change** can affect the size, shape or color of a substance. The original substance stays the same, no new substance is formed.

Hypothesis: Do you think that there will be a taller geyser from the regular or the diet Coke? Why or why not?

____

____

Materials:

2 Cups	**Bottle Coke, 16.9 oz – Room Temp.**	**Magnifying Glass/Microscope**	**Measuring Tape**
2 Index Cards	**Bottle Diet Coke, 16.9 oz – Room Temp.**	**Package Mint Mentos**	

Procedure:

1. Use your magnifying glass or pocket microscope to observe the exterior of a Mento. Write down your observations.
2. Label your cups: Coke and Diet Coke. Pour a small amount of each type of Coke in the cups.
3. Place one Mento in each cup and observe what happens. Write down your observations, noting any differences or similarities between the two types of sodas.
4. If you are not outside yet, it is a good idea to go outside now. Set up the measuring tape against a wall so that you can measure how high the geyser shoots from each of your bottles of Coke.
5. Place one of your bottles of Coke in front of the measuring tape, on a level surface.
6. Roll your paper to make a tube. Line up 5 of your Mentos in the tube, in a stack.
7. Cover the bottom of the tube with another index card and place it over your bottle of soda.
8. Make sure that your tube is right over the top of your bottle so that the Mentos will drop in once the index card is removed. Now, quickly slide the index card out so that the Mentos all drop into the Coke at once.
9. Get out of the way! This is a wet zone, so expect to get wet!
10. Observe and record how high the geyser reached on your measuring tape and then repeat steps 5-9 with your other Coke.

Data Tables:

Qualitative Data:

Qualitative data describes something. In an experiment, it is what you can see or notice but cannot measure in numbers. (Hint: think quality for description.)

	Observations
Mentos	
Diet Coke	
Regular Coke	
Mentos + Diet Coke	
Mentos + Regular Coke	

Quantitative Data:

Quantitative data can be counted or measured. It can be measured in numbers. (Hint: think quantity for numbers.)

	Height of Geyser
Diet Coke	
Regular Coke	

Graph:

Create a bar graph to compare how high the geyser of each type of Coke was. In science, it is always better to use the metric system when measuring. But since many United States measuring tapes are only in inches, it is okay if you used inches to measure for this lab.

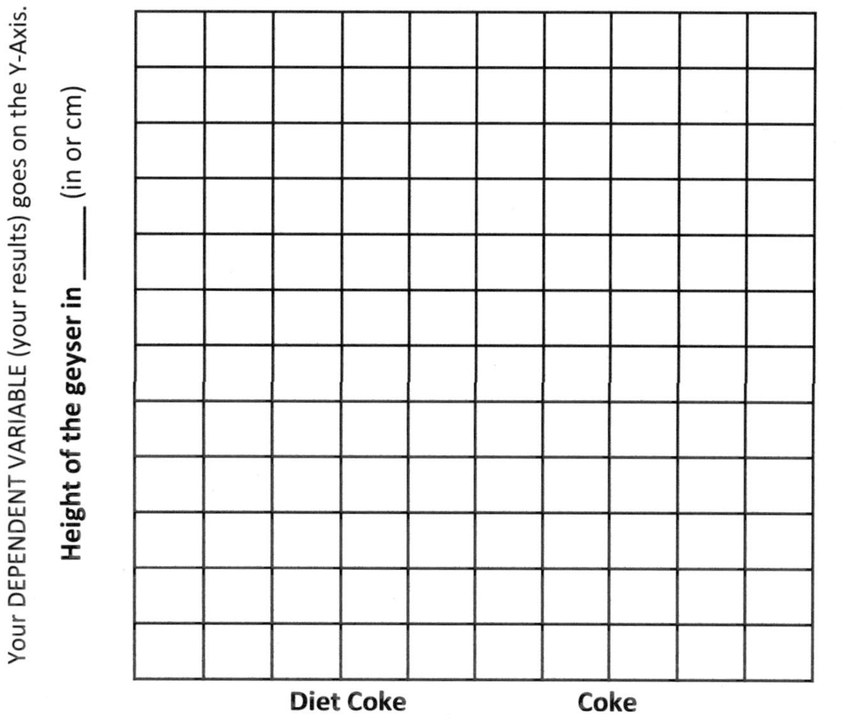

Your INDEPENDENT VARIABLE (what you are testing) goes on the X-Axis.

Conclusion:

On another sheet of paper, use your data and observations to write whether your hypothesis was supported or not supported. By the way, an unsupported (not proven) hypothesis is perfectly okay! Discuss why you think you got the results you did. Also, include what you might have done better and another experiment you might do based upon this lab.

What is Chemistry?

Name:

There are so many fantastic free learning videos on the internet. Answer the questions on this sheet as you watch the following videos. You might need to re-watch these short videos once or twice, to answer all the questions.

Importance of Chemistry in Life (3:03 min)

Go to: tinyurl.com/ykd5587v

1. When we eat breathe or just sit down what is taking place?
2. Chemistry is the study of ________________.
3. What are 3 examples of chemical reactions?
4. What are 3 important uses of chemistry?
5. What is something that people used to die from?
6. What can studying chemicals help us to do?

Chemistry Life Hacks (3:47 min)

Go to: tinyurl.com/3k2ppy6w

1. What is built up in an egg as it goes bad?
2. What about an onion reacts to make your eyes tear? What is the fix?
3. What gets replaced in veggies, causing them to lose their bright green color?

Importance of Chemistry in Life (3:03 min)

Go to: tinyurl.com/3zys6fbv

1. What is the most exciting phrase to hear in science?
2. What were the chemicals in coal tar similar to?
3. Instead of Aspirin what did Perkin end up discovering? Why was this a big deal?
4. Because of Perkin what sprang up trying to make treasure out of trash? How is this a good thing?
5. What did Fahlberg discover by not washing his hands?
6. Fluorinated hydrocarbons led Plunkett to discover ________________. What was it good for?
7. Because Teflon didn't stretch well. Bob's frustration allowed him to discover what?

Got More Time? Extra Video: Chemical Curiosities: Surprising Science (1:09:42)

Go to: tinyurl.com/2p8mpr4r

Watch these fantastic demonstrations from the Royal Institution. For each demo you watch, **write down what the demo was and the science behind it**. You can do this on the back of this paper if it is blank or on another sheet of paper.

Scientific Method Experiment Challenge

Name:

You are going to use the scientific method as you design a simple experiment. You will test the experiment yourself. The purpose of this assignment is to give you practice when experimenting. Pick something simple to test that you can finish in 1-2 days max! If you are a little confused about the scientific method, try watching this Science Jam video: tinyurl.com/7ukupju

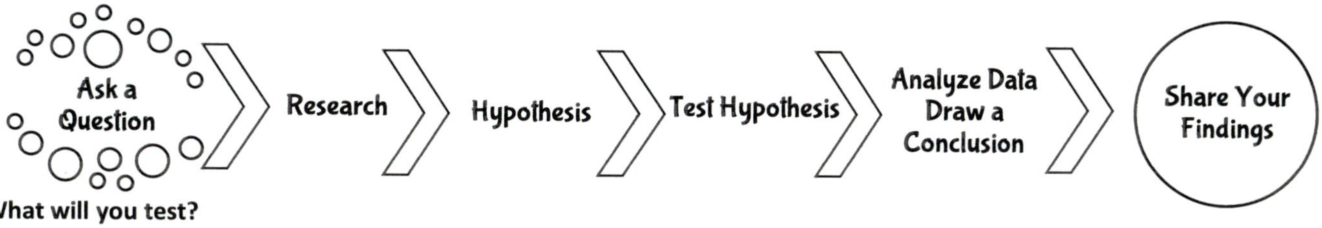

What will you test?

There are many simple questions you could design an experiment around. If you still have extra coke and Mentos you could even further explore this topic. Do more Mentos make a larger geyser? Do other diet sodas work just as well? Here are some other simple ideas you could test, but do not feel like you have to stick to this list. If you are wondering about something else than go ahead and explore it with permission from your teacher or parent!

Simple Ideas

- Which side of a penny holds more drops of water?
- What style of paper airplane will fly the farthest?
- Which type of bubble gum has the longest lasting flavor?
- Can you add something to ice water to make it colder?

List your Question on the lines below:

____

____

Parent/Teacher Initials

Variables:

Whenever you complete an experiment, you have a variable that you test or change on purpose, which is the independent variable. You will also have a variable you will measure to see your results: the dependent variable. The control group is to see if what you are testing makes a difference. All the other variables must be kept the same in order for the experiment to be fair, which are the controlled variables.

	Example from Mentos Diet Coke Experiment	Your Variables
Independent Variable	Regular Coke	
Dependent Variable	Height of the Geyser	
Control Group	Diet Coke	
Controlled Variables	Same amount of Coke Same number of Mentos Measuring tool used the same way	

Background Research:

Look up information on your topic. Make sure to read at least two sources. Write what you have learned in the space below.

Hypothesis:

Now that you have some background knowledge, write down what you think the results of your experiment will be.

____

____

Parent/Teacher Initials

Materials:

Procedure:

Write a detailed procedure in the lines below. Use as many lines as necessary.:

1.
2.
3.
4.
5.
6.
7.
8.
9.
10.
11.
12.
13.
12.
13.
14.

Data Tables:

BEFORE beginning your experiment, draw your data tables, on another paper. It is always important to know how you will collect your data ahead of time. It is also important to collect data in an organized way so that you do not get confused.

Parent/Teacher Initials

Graph:

Compare your data by creating a graph, on another sheet of paper. Make sure you label your graph correctly.

Conclusion:

On another sheet of paper use your data and observations, to write down whether your hypothesis was supported or not supported. Discuss why you think you got the results you did. Also, include what you might have done better and another experiment you might do based upon this lab.

Parent/Teacher Initials

Matter & Its States

States of Matter Student Notes Name:

Your teacher or parent will give you these notes or have you go to the resource page to watch the recorded presentation.

Resource page: tinyurl.com/2m5fcyfs (password: Matter).

Matter

- Takes up ___________, and has ___________.
- ___________ states of matter
- Made up of ___________ that are ___________ in motion. ___________ down as they get ___________
- ___________ is determined by how much ___________ a particle has (___________ energy)

Solid

Draw Particles:

- Have a ___________ ___________
- Have a ___________ ___________
- Arranged in ___________ regular ___________ and move very ___________.
- ___________ packed

Liquid

Draw Particles:

- Have a ___________ volume but ___________ definite ___________.
- ___________ tightly ___________ than solids.
- ___________ ___________ past each other and change ___________ to fill ___________.

Gas

Draw Particles:

- Has no ___________ ___________.
- Has ___________ fixed ___________.
- Particles move ___________.
- ___________ ___________ between particles.

Plasma

Draw Particles:

- Is ___________ gas.
- ___________ electrons stream through ___________ ___________.
- Has ___________ fixed ___________ or ___________.
- Conducts ___________ and is affected by ___________ ___________.
- ___________ of universe is composed of ___________!

Watching Ice Melt

Name:

This might seem like a strange lab to do, BUT it will teach you something important about temperature and states of matter.

Goggles and Aprons on!
You will be working around a heat source and boiling water.
This lab should only be done under adult supervision.

Question: How will increasing the heat affect the state of H_2O?

Background Research: The first law of thermodynamics states that energy can be transformed, but cannot be created or destroyed. This leads into the concept of how different substances can change from one phase to another by absorbing or releasing energy. In an **endothermic** phase change a substance gains heat energy. In an **exothermic** phase change a substance releases energy into the environment. When the system is heated, it changes by increasing its temperature. A plot of the temperature versus time is called the heating curve.

Hypothesis: Use what you know about water to write your hypothesis.

____

____

Materials:

Ice	**Beaker or Pot**	**Thermometer**	**Heat Source: Stove, Hot Plate, Bunsen Burner, etc**
Timer (can use phone)	**Cup**		

Imagine you put the following substances in hot water. Predict what will happen to them.

Substance	In Hot Water it Will...
Butter	
Rock	
Chocolate	
Dry Ice	

Intro Activity: Imagine you put an ice cube in a cup of lukewarm water. What do you think will happen to the ice cube? What do you think will happen to the temperature of the water?

1. Fill a cup or beaker half full with lukewarm water then measure and record the water temperature under beginning temperature for Ice Cube #1.
2. Set a timer as you add an ice cube to the water. Gently stir it with your thermometer until the ice cube melts. Now record the time it took for the ice cube to melt and the temperature of the water under Ice Cube #1.
3. Immediately add a second ice cube to the water. Set your timer and then stir this ice cube until it melts. Now record the melting time and ending temperature of the water as soon as it melts.

Ice Cube Number	Beginning Temperature	Ending Temperature	Melting Time
1			
2			

4. Which of the ice cubes melted fastest? Why do you think it took longer for one of the ice cubes to melt?

5. Do you think if you put an ice cube in hotter water that it would melt at a different rate than if you put the ice cube in colder water?

Procedure

1. Fill your small pot or glass beaker with ice.
2. Suspend the thermometer in your container so that it does not touch the bottom. Depending upon your thermometer there are a few ways to do this. If your thermometer already has a clip you can use that. Otherwise you can use a thermometer clamp or a clothes pin.
3. Record the starting temperature of your ice and the state it is in (S:Solid, L:Liquid, G:Gas) in your data table.
4. You will now turn on your heat source and start your timer. For a stove or a hotplate, you would turn it on high.
5. Every 30 seconds you will record the state and temperature of your container.
6. Continue until your temperature has stopped changing at the boiling point.

Data Table:

Time (min.)	S/L/G	Temp: °C or °F	Time (min.)	S/L/G	Temp: °C or °F	Time (min.)	S/L/G	Temp: °C or °F
0.0								
0.5								
1.0								

Graph

Create a line graph of your data that includes the following:

- Label the x-axis with the time (in minutes).
- Label the y-axis as temperature (in degrees either Fahrenheit or Celsius).
- Create a title for your graph that describes the data.
- Color your line to show what state your water was in. Green while it was a solid, orange as the ice was melting, blue for liquid and yellow for water boiling as it is turning into a gas. Also label the following points on your graph: MP (Melting Point) and BP (Boiling Point).

Conclusion and Analysis:

1. How did adding heat energy affect the state of your water?

2. Were the phase changes you observed today exothermic or endothermic?

3. Explain what was happening to the water molecules in the flat areas of your graph.

4. Explain what was happening to the water molecules in the sloped areas of your graph.

5. If you were adding thermal energy the whole time, why were there times when the temperature did not change much? (What was the thermal energy doing those times?)

6. When the water was boiling and becoming a gas, was it releasing energy or absorbing energy? Explain your answer.

7. If you put liquid water in the freezer would it be absorbing or releasing heat as it froze? Explain your answer.

8. Look at this graph of a waxy material called lauric acid. At what temperature does lauric acid change from a solid to a liquid? Explain your answer.

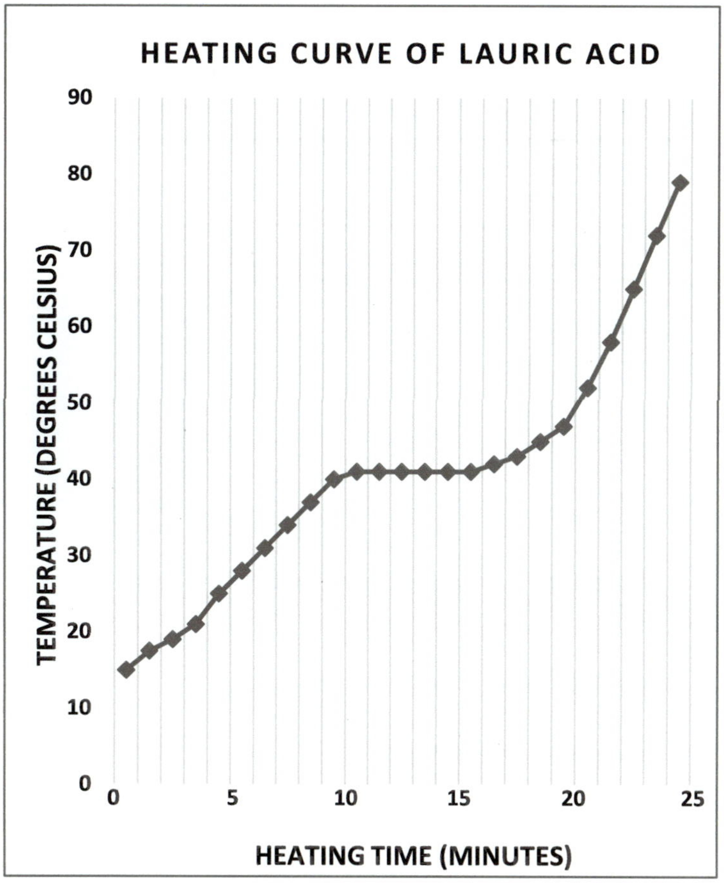

Matter Labs

Name:

There are so many fantastic free activities and interactive labs online. Answer the questions on this sheet as you explore these activities and online labs.

Basic Solid, Liquid and Gas Exploration

Go to: tinyurl.com/2sz89r8m

This interactive presentation lets you try to compress and change the shapes of solids, liquids and gases.

1. In a solid how are particles held together?

2. In a liquid the particles are still held together but what are they able to do?

	Solid	Liquid	Gas
Could you change its shape?			
Why?			
Could you compress it?			
Why?			
3 Examples of the State			
Why are these properties useful?			

Properties of Matter Study Jam

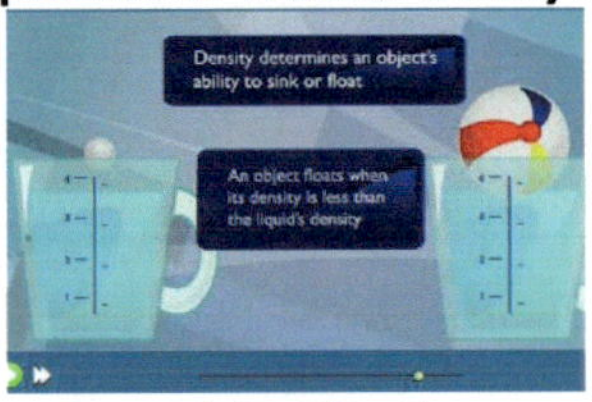

Go to: tinyurl.com/7f5t8l3 -Choose play video.

1. Why is a golf ball heavier than a much larger beach ball?

2. What is Density? What is its formula?

3. List 4 properties of Matter:

4. Define Mass.

5. How are Mass and Weight different?

6. Define Volume.

7. What is the trick for measuring the volume of things that cannot be measured with a ruler?

8. Why does a golf ball sink in water while a beach ball floats?

States of Matter Interactive

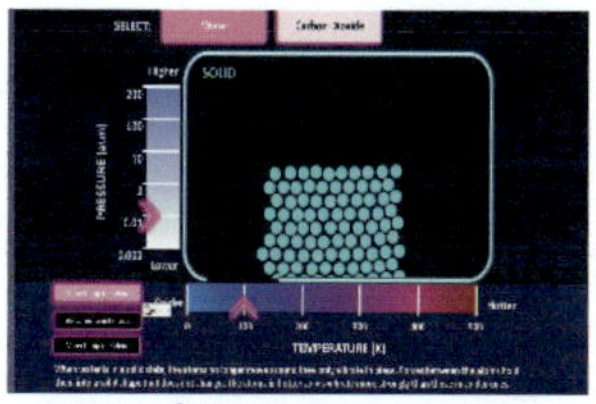

Go to: tinyurl.com/2hn5thzp – Launch Interactive

Water is the only substance on Earth that occurs naturally as a solid, liquid and a gas. Scientists can artificially transform other substances by manipulating their temperature and pressures. In this lab you will get to play with 2 substances: Water and Carbon Dioxide to see how you can change their particle movement and states.

1. Select Water and slide the pressure up to 1 atm.
2. Select to "View temp in Fahrenheit".
3. Slide the temperature scale until the water enters a liquid state. Write this temperature in Fahrenheit _____ and in Celsius _____
4. Continue to slide the scale until the water enters a gaseous state. Write this temperature in Fahrenheit _____ and in Celsius _____
5. Without touching the temperature can you change this water back to a liquid or a solid by adjusting the pressure? Explain what you did why you think you got your results.

6. Fill out this data table with the state of the particles at each temperature. Adjust the pressure to 1 atm.

Water					
Carbon Dioxide					
Temperature	-279.7 F -173 C	-99.7 F -73 C	80.3 F 27 C	260.3 F 127 C	440.3 F 227 C

7. At 0.01 atm is there a difference in the Carbon Dioxide atoms when you change the temperature from 260.3 F to 440.3 F?

8. How would you say that temperature and pressure relate to the movement and state of a particle?

9. Can you find a way to make Carbon Dioxide a liquid? Can you come up with an explanation for this?

Evaporation or Sublimation?

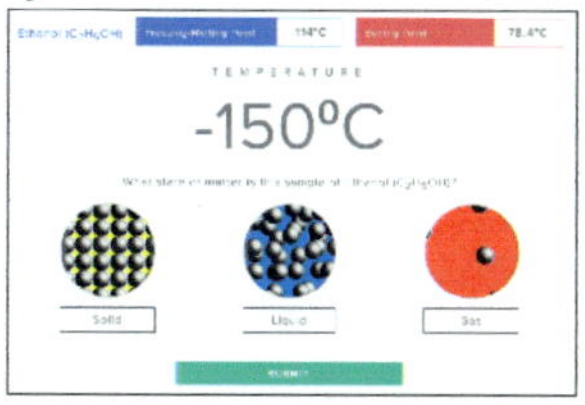

Go to: tinyurl.com/2weavurr

1. Figure out what state of matter each substance is in. **Hint:** At the top of each slide, read the freezing (solid) point and the boiling point (gas) to find the answer.

2. Use this guide when identifying the phase changes:
   - **Melting:** changing from a **solid** to a ⇨ **liquid**
   - **Freezing:** liquid ➡ solid
   - **Evaporation:** liquid ➡ gas
   - **Condensation:** gas ➡ liquid
   - **Sublimation:** solid ➡ gas
   - **Deposition:** gas ➡ solid

3. If you did not get a 10/10 start the quiz over until you do. Use the tips in steps 1 and 2 to help you get that perfect score!

☐ I got a 10/10!

Solid, Liquid, Gas Quick Lab

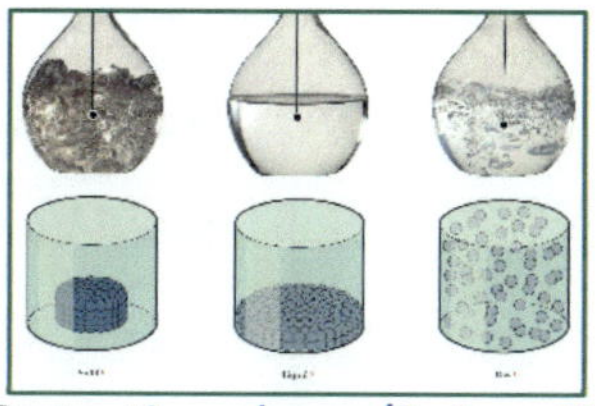

Go to: tinyurl.com/2u6vzw95

Draw and describe the particle movement for each state of matter:

Solid **Liquid** **Gas**

States of Matter – Mini Lapbook

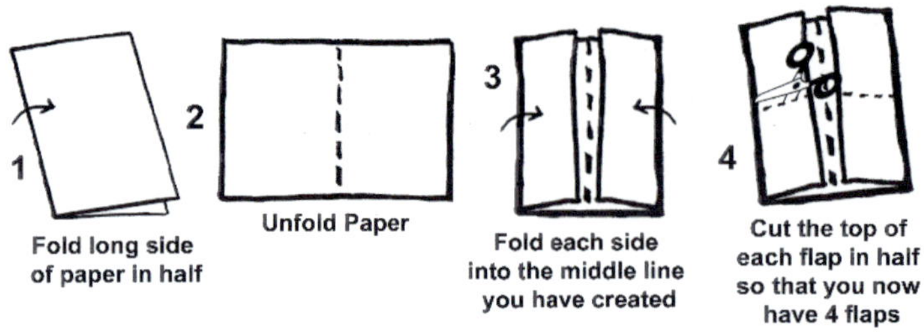

Take an 8.5 x 11 piece of computer paper. Follow these directions to make your lapbook folds.

Front

SOLID Definition in your own words. Draw a picture of the particles	**Liquid** Definition in your own words. Draw a picture of the particles
GAS Definition in your own words. Draw a picture of the particles	**PLASMA** Definition in your own words. Draw a picture of the particles

Look at the pictures on this page to see what information you need to include in your lapbook. Feel free to look up information you do not know. Just make sure to write your answers your own words! As you can see, each flap is dedicated to a State of Matter.

Place or draw energy arrows between the solid, liquid and gas on the inside middle of your foldable. Place an arrow showing how these states change into each other. You will use each of the following words:

Freezing	Melting
Boiling	Sublimation
Condensing	Evaporation

Color the arrows blue that show energy being lost and red where energy is being gained.

Inside Left	Inside Middle		Inside Right
Solid Vocab Crystalline Pressure Freezing Point	**SOLID** Draw or cut out picture examples.	**Liquid** Draw or cut out picture examples.	**Liquid Vocab** Viscosity Melting Point Condensation
Gas Vocab Compression Evaporation Sublimation	**GAS** Draw or cut out picture examples.	**PLASMA** Draw or cut out picture examples.	**Plasma Vocab** Ions Electrons Northern Lights

We cut 3 flaps into a post-it note to make this **Vocab Flip**. If you do not have post-its you can cut and tape a piece of paper. Write the definition of each word underneath its flap.

States of Matter – Mini Lapbook

You can cut and paste these images and words into your lapbook OR you can draw or find your own!

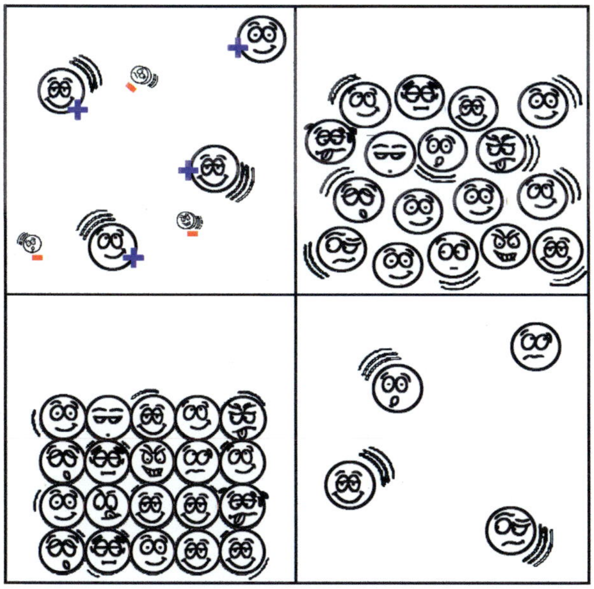

SOLID
LIQUID
GAS
PLASMA

Sublimation

Freezing

Boiling

Condensing

Melting

Deposition

Gas Laws & Properties

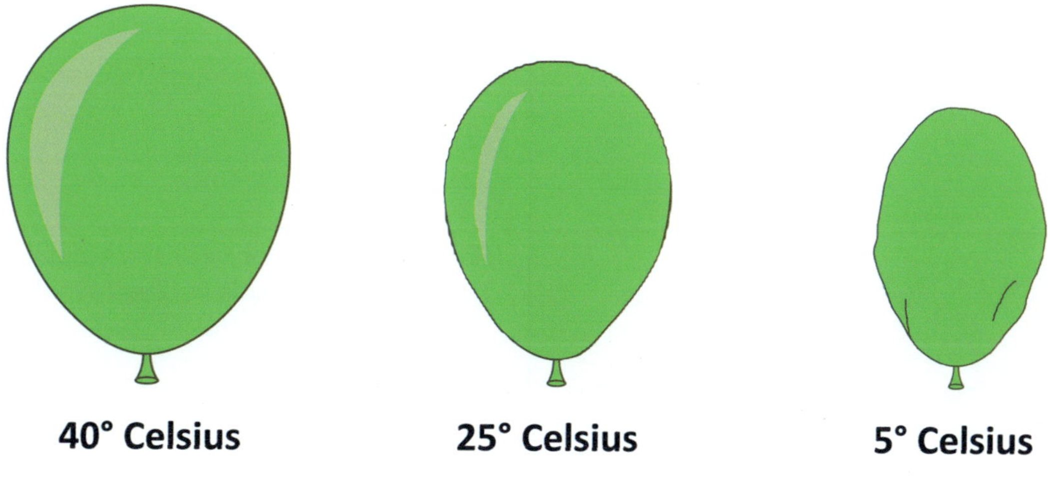

Gas Law Notes

Name:

Your teacher or parent will give you these notes or have you go to the resource page to watch the recorded presentation. Resource page: tinyurl.com/2m5fcyfs (password: Matter).

A Gas is one of the __________ states of matter. It has no __________ shape and no __________ volume. Its particles move so fast that they __________ and __________ around.

As ____________ increases the ____________ a ____________ moves.

Charles' Law

As __________ increases __________ increases.

T ____ = V ____

From your experience or the slideshow, draw a picture example to illustrate this law.

Scientific Law

Is:

Example:

Scientific Theory

Is:

Example:

Gay-Lussac's Law

As __________ increases __________ increases.

T ____ = V ____

From your experience or the slideshow, draw a picture example to illustrate this law.

Boyle's Law

As __________ increases __________ decreases.

T ____ = V ____

From your experience or the slideshow, draw a picture example to illustrate this law.

What law does ivory soap and the microwave show?

GAS LAW CHALLENGE LAB CARDS

Complete these mini lab cards to answer the questions on your "Gas Law Station Labs."

MARSHMALLOW SHRINKER (1)
BOYLE'S LAW

1. Put a mini marshmallow or three inside of your syringe and push the plunger down until it touches the top of your marshmallows.
2. Plug the tip of the syringe with its cap or by placing your finger on it.

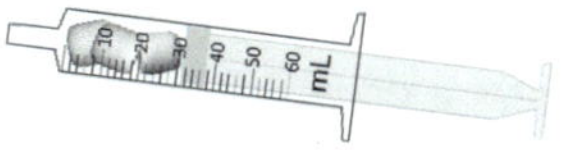

3. Now pull the plunger up. What happens to your marshmallow?
4. Release your finger or the cap from the tip and pull the plunger to the top of the syringe.
5. Plug the tip of the syringe again and push the plunger down. What happens to your marshmallow?
6. Which variables were at play in this lab: Temperature, Pressure and/or Volume?
7. Explain how this is demonstrating Boyle's Law.

FUN EXTRA: Use your syringe to demonstrate how an increased or decreased pressure can change the boiling point of a liquid. Fill your syringe with about 20 mL of warm water. Remove the air from the syringe by pushing a little on the plunger. Plug the tip with your finger or a cap and pull the plunger up. Do you see the water start to boil? Can you explain why this happens using what you know about pressure and particle movement/energy?

A BUNCH OF HOT AIR (2) - CHARLES' LAW

1. Take an empty 2-liter bottle or a gallon milk jug and put a party balloon on top of it.
2. Place your bottle in a bowl or casserole dish full of hot water, An adult can pour this for you. What happens to your balloon?
3. Now place your bottle in a bowl or casserole dish full of ice water. What happens to your balloon?
4. Which variables were at play in this lab: Temperature, Pressure and/or Volume?
5. Explain how this is demonstrating Charles' Law.

FUN EXTRA: If you used a 2-liter bottle for this lab, try making a Cartesian diver this shows **BOYLE'S LAW**, in action. Fill your bottle up with water. Put a ketchup or soy sauce packet in it (these have air trapped inside so they will float in the bottle). You can also trap air in a drinking straw by folding it and closing the ends with a paper clip. Screw the lid onto your bottle and squeeze the sides. What happens to the "diver" inside? Use what you know about Boyle's law to describe why you think this is happening.

CANDLE VACUUM (3) - GAY-LUSSAC'S LAW

SAFETY: An adult should be present. Wear your safety goggles.

1. Put water into a casserole dish or onto a plate, about ¼ inch high. If you have food coloring put 3-4 drops of a color into the water.
2. Place a tea or votive candle in the middle of the casserole dish.
3. Light the candle.
4. Place a clear cup or jar over the candle.
5. Watch as the candle goes out. What do you see happen to the water?
6. Repeat this experiment at least one more time, now that you know what you are looking at.
7. Which variables were at play in this lab: Temperature, Pressure and/or Volume?
8. Explain how this is demonstrating Gay-Lussac's Law.

FUN EXTRA: With adult supervision, put about a tablespoon of water in a clean soda can. Place this can on a hotplate or the burner of your stove. When you see steam coming out of the can, use tongs to pick up the can and place it upside down into a bowl of ice water. What happened? Use what you know about Gay-Lussac's law to describe why you think this happened.

*Note: If you do not have a tea candle you can use a votive or any candle that will fit under your jar or cup.

GAS LAW CHALLENGE LABS

Name:

Fill out the squares below from what you have already learned. Fill out the rest of this worksheet as you visit each station.

BOYLE'S LAW

Pressure and _______________ are inversely related. This means that if pressure goes s then volume goes r. If pressure goes _____ then volume goes s.

CHARLES' LAW

_______________ and volume are directly related. This means that temperature and volume either both go s or both go _____.

GAY-LUSSAC'S LAW

Temperature and pressure are directly related;. This means that the pressure and temperature either _______________ go s or both go r.

MARSHMALLOW SHRINKER - BOYLE'S LAW

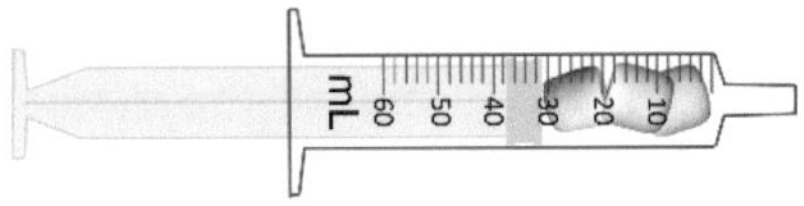

1. What happened to the marshmallows when you pulled the plunger up?

2. What happened to the marshmallows when you pushed the plunger down?

3. Circle the variables that were at play in this lab:

   Temperature **Volume** **Pressure**

4. How did this lab demonstrate Boyle's Law?

Boiling Water Extra: Why is the water boiling in your syringe?

A BUNCH OF HOT AIR- CHARLES' LAW

Safety: Wear your Goggles, Do this with Adult Supervision

1. What happens to the balloon in the hot water?

2. What happens to the balloon in the cold water?

3. Circle the variables that were at play in this lab:

   Temperature **Volume** **Pressure**

Charles' Lab Continued:

4. How did this lab demonstrate Charles' Law?

Ping Pong Extra: What happens to the ping pong ball? Why?

BOYLE'S LAW – Cartesian Diver Extra: What happened to your diver? Using what you know about Boyle's law describe why you think this happened.

CANDLE VACUUM- GAY-LUSSAC'S LAW

Safety: Wear your Goggles.
Do this with Adult Supervision.

1. Draw a picture showing what happened to the water as the candle went out?

2. Circle the variables that were at play in this lab:

   Temperature **Volume** **Pressure**

3. How did this lab demonstrate Gay-Lussac's Law?

Crushing Can Extra: What happened to the can? Use what you know about Gay Lussac's Law to describe why this happened.

Simulators and Syringes

Name:

There are so many fantastic free activities and interactive labs online. Answer the questions on this sheet as you explore these activities and online labs.

Interactive Syringe

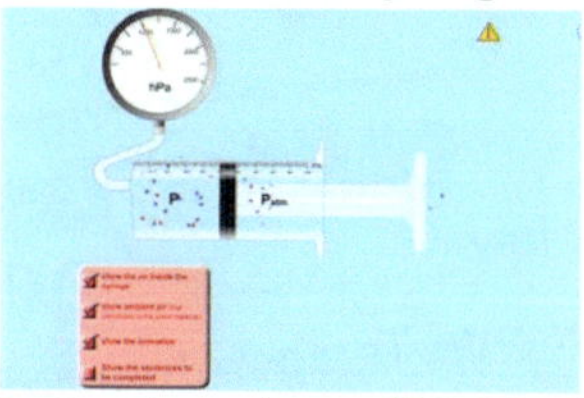

Go to: tinyurl.com/54ah9sah

Accept the cookies (it will ask for acceptance in French) and then click the play button to enter. Check all but the bottom box so that you can see the inside and outside air.

1. Can you push the piston of the syringe all the way to zero? Why or why not?

2. How do you make the pressure higher (hPa: hectopascal is the SI unit for measuring pressure)? What is the highest you can get it to go?

3. How do you make the pressure lower? What is the lowest the pressure will go?

4. So when the Volume of the air in the syringe decreases, the molecules get _____________ together and the _____________ increases. What law does this represent?

Phet: Gas Particle Simulator

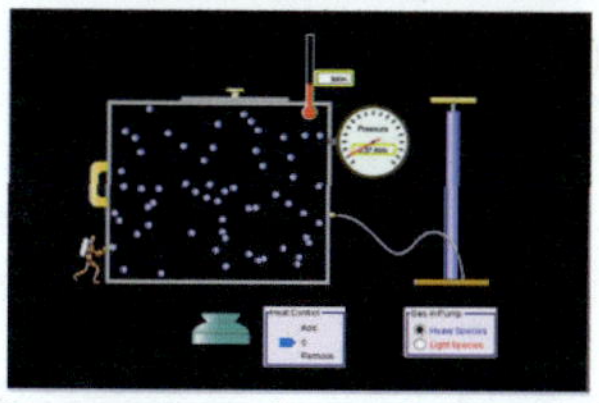

Go to: tinyurl.com/yckk4d2b

Under Gas Properties choose **Ideal**

1. Take a few minutes to play around with this simulation. Can you change the pressure, temperature and number of molecules in your tank?

2. Can you blow the lid off the container just by changing the mass (amount of gas molecules) in the container?

3. Hit reset and pump air 3 or 4 times into the container. What happens to the particles when you add or remove heat to the container?

4. Describe how changing the volume affects pressure and temperature in this simulation.

5. Describe how changing the temperature affects pressure and volume in this simulation.

6. Make the temperature constant (stay the same) by selecting it in the upper left. How are pressure and volume related in this simulation?

7. Fill in the table below. Hit reset before you begin working on a new law. Don't forget you can freeze or make each item constant on the left side.

	Charles' Law	**Gay-Lussac's Law**	**Boyle's Law**
How did you demonstrate this in the simulation?			
Volume: decrease/ increase			
Temperature			
Pressure Atm			

8. Use the simulator to solve this question:
You check the pressure of your bike's tires on a very cold day and find that they are 30 psi (psi = pounds per square inch. It is another way to measure pressure). You filled your tires just the day before It was very warm that day. Assuming there is no leak in your tires, what would explain this change in pressure? What law is this?

A Real Case of Gas Laws

Name:

Different scientists helped to develop our Gas Laws. Use your notes and what you have learned about each gas law to fill in the blanks for each of the three laws below. Then tell us which law matches with the real-world examples below.

Which Law? Charles, Boyle, Gay-Lussac:

1. ______________ Pushing the nozzle of a can of air freshener opens up a hole in the can which allows the air freshener expand out of the can.
2. ______________ You take your basketball out of the cold garage and it is too flat to play with. The next day is warm and the basketball is at full pressure and bouncy.
3. ______________ When you open a hot oven the hot air inside puffs out at you. This is because the air is under a higher pressure in the oven.
4. ______________ A torch heats the molecules inside of a hot air balloon. The balloon inflates and carries the people inside of the basket up into the air.
5. ______________ When gunpowder burns, it creates a large amount of superheated gas. The high pressure of the hot gas behind the bullet forces it out of the barrel of the gun.
6. ______________ Placing a balloon in extremely cold liquid nitrogen makes the balloon much smaller. When the balloon is put back at room temperature it gains its volume back.
7. ______________ The thermometer inside of the Thanksgiving turkey pops up when the temperature inside of the bird is high enough for it to be safe to eat.
8. ______________ Twist a straw together on both sides until there is a tight pocket of air. When you flick this middle part of the straw with your finger, the straw bursts open.
9. ______________ Put a hot can, with hot air inside of it, upside down in cold water and the can will crush in upon itself.
10. ______________ Since there is no atmospheric pressure in space, a spacesuit is vital. It keeps the gases in an astronaut's blood stream from expanding and killing them.
11. ______________ The can of PAM I use to spray my pans before cooking, says to not get it close to a heat source. This is because if the compressed gas in this can gets too hot, the can will explode.
12. ______________ The bends is something all scuba divers want to avoid. With the extra pressure of water on top of a diver, the nitrogen gas decreases in volume and gets absorbed into the diver's blood cells. If the diver ascends too quickly, the nitrogen will rapidly expand, causing the cells in the diver's body to rupture. This can be super painful and even deadly.

Charles' Law

In this law ______________ stays constant or the same. Temperature and ______________ are directly related. This means that the temperature and volume either both go ______________ or both go up.

Boyle's Law

In this law ______________ stays constant or the same. ______________ and Volume are inversely related. This means that if pressure goes down then volume goes up, and if pressure goes ______________ then volume goes down.

Gay-Lussac's Law

In this law ______________ stays constant or the same. Temperature and pressure are directly related. This means that pressure and temperature either ______________ go down or both go up.

Mixtures & Solutions

Solution to a Great Mix Up!

Name:

There are many fantastic free learning videos on the internet. Answer the questions on this sheet as you watch the following videos. You might need to rewatch these short videos once or twice to answer the questions.

The Great Picnic Mix Up: Crash Course Kids (4:10)

Go to: tinyurl.com/2s45ncjy

If this is too fast, you can slow down the video. Go to settings ⚙ > Speed> 0.75

1. Anytime you combine ____________ or more different things you get a ______________.
2. Circle the solution:
   Sand and Water   Sugar and Water
3. The SOLVENT is ____ and the SOLUTE is ____.
   a. The part of the solution that is dissolved
   b. The part which does the dissolving
4. What is solubility?
5. What is saturation?
7. Put and **M** next to the mixture and an **S** next to the solution at the picnic.

   Iced Tea _____   Fruit Salad _____

Mixtures and Solutions: Brainstew (23:14)

Go to: tinyurl.com/5aj4brcs

Into what two groups can we divide the states of matter?

1. What happens if you let salad dressing sit for a while?
2. Why is her breakfast cereal a mixture?
3. Why is the powdered drink she makes a solution?
5. All ______________ are ______________ but not all ______________ are ______________.

6. How did they separate the mixture of charcoal and water in the lab?
7. What are the two white solutions they test against each other? Which one wasn't a mixture and how do you know?
8. As they add sugar and vanilla to milk and cream, they are turning it into ice cream. Ice Cream is a ______________.
9. How many times a day do they have to milk a cow?
10. Milk right out of a cow is a ______________. Milk from the store is a ____________.

Dirty Laundry Crash Course (stop at 7:00)

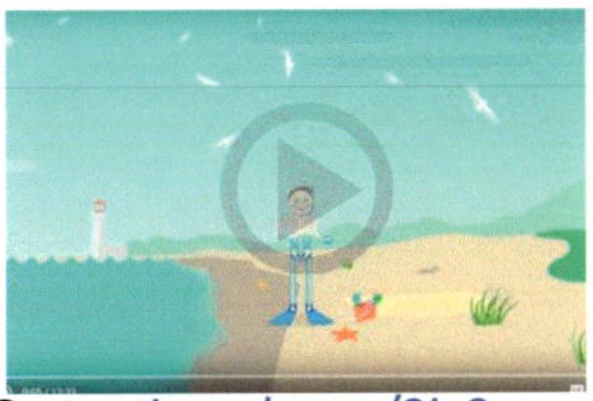

Go to: tinyurl.com/2jv9wcwx

This talks about the universal solvent (water). This is a little advanced, but he is fun to listen to. For students who want to know more, it is a great episode!

1. Without water, none of the ___________ ______________ that keep you alive would happen.
2. Why is water the key?
3. Why is bleach so good at its job?
4. Hydrogen peroxide is a ________. Water is the __________ and hydrogen peroxide the __________.
5. Why is water such a good solvent?
   a. There is a lot of it.
   b. It is ________________
   c. It is ________________
6. A water-based solution is an ______________ solution.
7. What is something that water does not dissolve?

Fake Blood Lab

Name:

We are going to make edible stage blood as we learn about solutions, solutes, solvents, viscosity and saturation!

Aprons on!
This can be a messy lab so dress appropriately. This is a good lab to do outside or in an area that can be easily cleaned up.

Background Info:

- In this lab, you will dissolve solutes like sugar, corn starch and cocoa powder into a solvent (water). Water is considered the universal solvent because it is a liquid at a wide range of temperatures and it can dissolve so many different things.
- We are adding 4 times the amount of powdered sugar to our solvent than we have water. In doing this we are creating a thick super saturated solution.

Materials:

Water	**Ziploc Bag**	**Corn Starch**	**Red Food Coloring**	**Clear Cup (greater than ½ cup capacity)**
Teaspoon	**Spoon**	**¼ Measuring Cup**	**Cocoa Powder**	**1 Cup Powdered Sugar**

Procedure:

1. Measure ¼ cup of water into your cup. This water is your solvent, observe its **viscosity** (flow/thickness) and **clarity** (clearness).
2. You have four times the amount of solute (sugar) as you do solvent (water). Do you think that the entire cup of powdered sugar will dissolve in this water? Mark on the cup how high you think your solution will rise, in the cup, once you have mixed all the solute and solvent.
3. Use your spoon to slowly sift powdered sugar into your cup, occasionally stirring the solution with the spoon. Is the water still clear? Has the viscosity changed?
4. Pour your sugar/water solution into your baggy. Make sure to firmly seal the top. Use your fingers to finish mixing this solution through the bag.
5. Add 25 drops of red food coloring. Has this affected the clarity or viscosity of your solutions?
6. Add a teaspoon of cocoa powder. When something dissolves easily in water it has a **high solubility**. Comparing this cocoa powder to the powdered sugar, which had a higher solubility?
7. In your cup mix 1 teaspoon of water with 2 teaspoons of corn starch until this mixture is smooth and free of lumps. Now add this cornstarch mixture to your bag.
8. Observe your fake blood. Would you say that it is a solution or a mixture? Why? How has the viscosity and clarity changed since you started out with just water?

Data Tables:

Qualitative Data:

These solutes are made of different molecules and interact with water in their own unique ways. List the similarities and differences between how these solutes mixed with the solvent (water).

Cocoa	Sugar	Cornstarch
	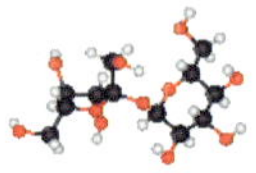	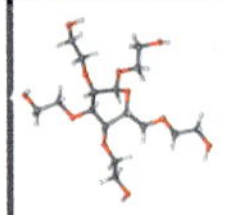

Questions:

1. In your own words describe:
   a. Solute:

   b. Solvent:

   c. Solution:

2. List 2 things from the real world with a high viscosity and two things with a low viscosity.

3. Were you surprised about how much sugar you were able to dissolve into your water and at how little the water rose? Use the pictures to help you explain why adding ¼ cup water with 1 cup sugar did not equal 1 ¼ cups of solution.

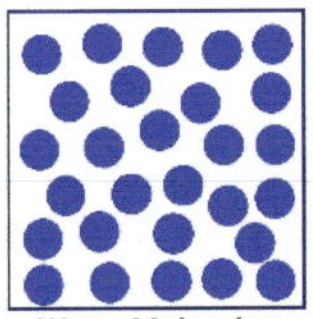

Water Molecules

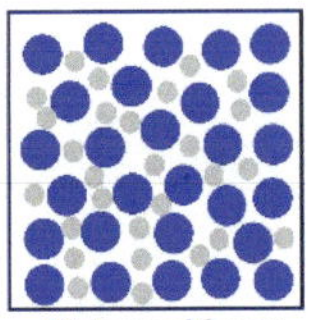

Water and Sugar Molecules

Fun Extra: Edible Barf Mixture

Nasa chemists actually simulate the texture and smell of real vomit to test their barf bags. We are just going to make a mixture that LOOKS like vomit. Check out this video if you want to see real Nasa chemists working in their lab to make simulated (fake) vomit! tinyurl.com/3br6zy6j

To make realistic looking but edible barf follow this recipe. Why is this a mixture and not a solution?

1 pkg unflavored gelatin	**¼ cup applesauce**	**yellow food coloring**	**¼ tsp cocoa powder**
¼ cup oatmeal	**¼ cup raisin bran cereal**	**microwavable bowl or mixing cup**	

1. Heat the applesauce in the microwave for 1 minute. This will be hot so be careful and have an adult present.
2. Stir in 1 packet of unflavored gelatin for 1 minute.
3. Add 2 drops of yellow food coloring and the cocoa powder. Mix thoroughly.
4. Mix the oatmeal and raisin bran together. Add half of it into your applesauce mixture and stir very gently just to mix.
5. Spread this barf mixture onto a plate so that it resembles vomit, adding extra raisins or cereal pieces as desired.
6. Allow the barf to cool completely (a couple of hours). Remove it from the plate with a spatula.
7. Have fun!

Great Ways to Get Together

T-Chart

A T-Chart is used to compare two different concepts or things. Today you are going to make a T-Chart comparing Mixtures and Solutions as you read an online article by Chem4Kids: tinyurl.com/4mndk5p .

There are four short pages to read in this article: Mixture I, Mixture II, Solutions I and Solutions II. As you read, make your T-Chart on another sheet of paper, following these guidelines.

Mixtures	Solutions
• Define what a Mixture is.	• Define what a Solution is.
• Does a chemical or physical change occur?	• Does a chemical or physical change occur?
• Heterogenous and/or Homogenous? Why?	• Heterogenous and/or Homogenous? Why?
• Fact/Difference	• Fact/Difference
• Fact/Difference	• Fact/Difference
• Examples	• Examples
• Picture	• Picture

Mixtures and Solutions T-Chart

Name:

Mixtures	Solutions

Atom, Molecule or Mixture?

Name:

Atoms and Molecules are words you need to know if you are going to deal with mixtures and chemistry. Each atom is the basic unit of a chemical element. A molecule is when two or more different or the same types of atoms, chemically join together. See if you can identify the atoms, molecules, and mixtures of them below. Write the correct label under each square. You will use some of them more than once.

Atoms	**Molecules**	**Mixture of Atoms**	**Mixture of Molecules**	**Mixture of Atoms and Molecules**

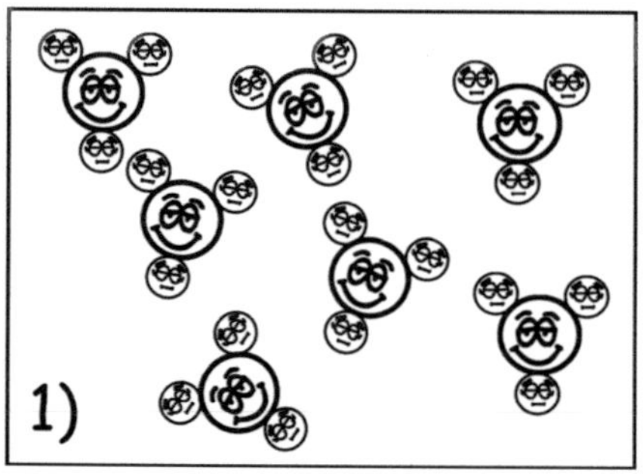
1) ______

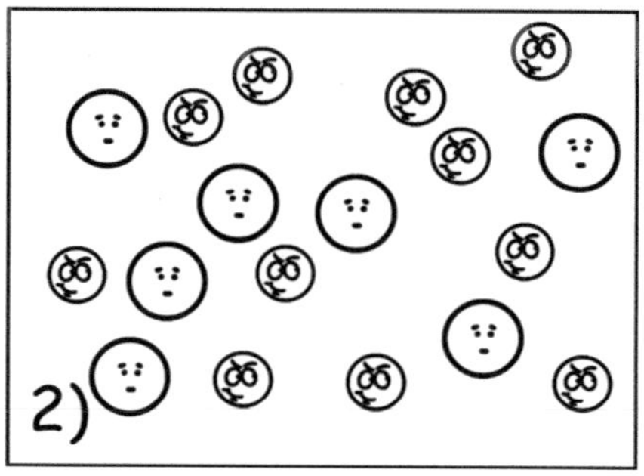
2) ______

3) ______

4) ______

5) ______

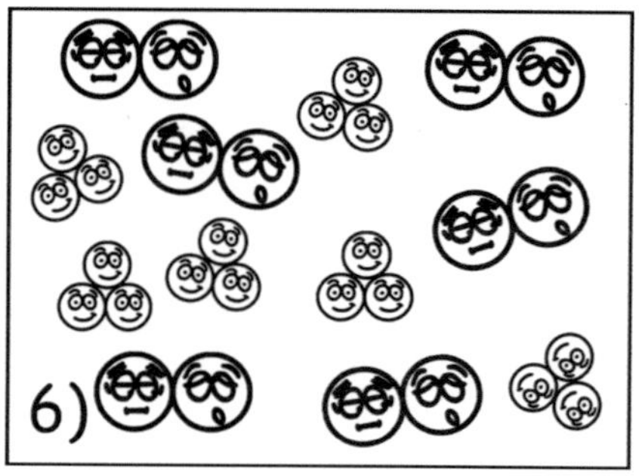
6) ______

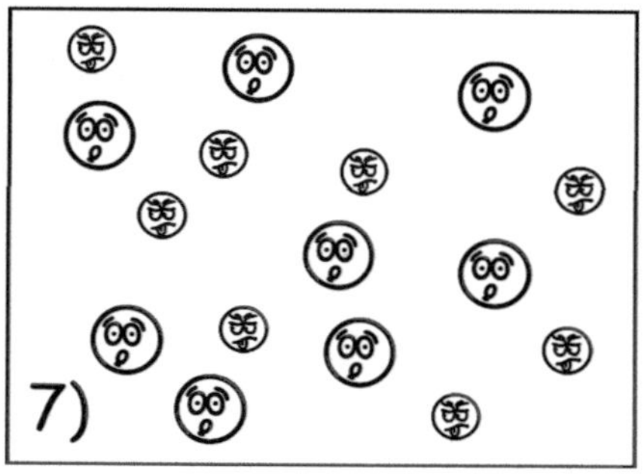
7) ______

8) ______

9) ______

______ **10)** A **compound** is a molecule that contains at least two different types of atoms. All compounds are molecules but not all molecules are compounds. Write down the number of each box that contains a compound.

______ **11)** An **element** is a substance that is made entirely from one type of atom. Write down the number of the boxes that are elements.

Atoms & Atomic Structure

ATOMIC STRUCTURE INTERACTIVE NOTES

Name:

Chemistry is really all about how atoms behave and interact with each other. We will take a look at the parts of the atom and learn about the properties of these atomic pieces. Go to this site to launch your interactive tutorial on Atomic Structure. tinyurl.com/sfweatfj

Fill in the blanks as you learn about the three distinct particles that make up an atom. Draw a line from each square to the particle on the picture of the carbon atom that it is talking about.

PROTON

- Found in the ______________.
- The ______________ of protons determines the ______________.
- #Protons = __________ __________
- It is a ______________ piece of the atom.
- ______ Charge.

ELECTRON

- ______________ part of the atom.
- Virtually ________ mass
- Always ______________ around the nucleus.
- ______ Charge.
- Electrons are involved in all ___________ ___________

NEUTRON

- Found in the ______________.
- ______ Charge.
- Holds the ____________ together.
- It is a ______________ piece of the atom, similar in mass to the proton.

Write down the number of each atomic particle in this carbon atom.

Protons: ________

Electrons: ________

Neutrons: ________

Fill in the properties of each subatomic particle as you answer these questions in your tutorial.

	Proton	Electron	Neutron
charge			
mass			
location			
importance			

Rainbow Flame Lab

Name:

We are going to heat different compounds and elements with fire to see what color light they produce.

Goggles on! Tie hair back and do not wear loose clothes, long sleeves or jewelry. This lab involves fire and should be performed with a supervising adult. Use caution and have appropriate safety equipment nearby.

Background Research:

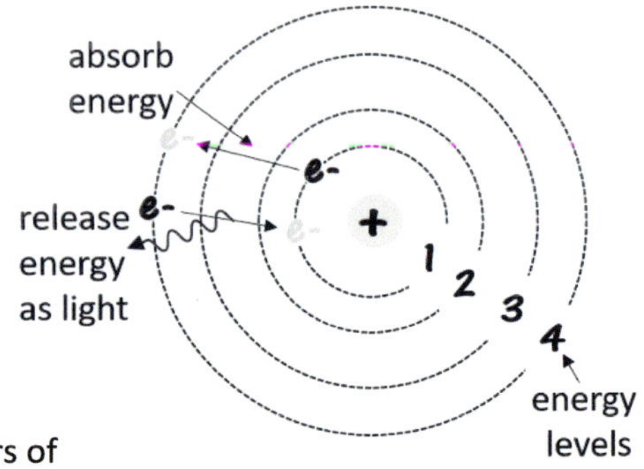

- An atom's electrons stay in **energy levels**. Electrons stay in the lowest energy level they can. Here they are stable and happy, this is called the **ground state**.
- When energy such as fire (heat) is added, the electrons get excited (**excited state**) and jump up levels.
- After being excited they then fall back down to their normal energy levels, releasing energy as **light**.
- Atoms of the metal elements release energy uniquely and different colors of visible light are produced. **The color of this light can be used to identify these elements,** especially when using a spectroscope or prism**.**
- The ability of metal atoms to produce these colors is used in the making of **fireworks**. By including different metal salts, or mixtures of metal salts, firework manufactures can produce beautiful displays in nearly all the colors of the rainbow.
- These different **colors produced by elements** are also how we **identify what makes a star**. Since we obviously can't visit or send equipment to such an incredibly hot place, we look at the colors of the light spectra coming from stars to identify what makes them.

Materials:

Sodium Chloride (Table Salt)	**Potassium Chloride (Salt Substitute)**	**Copper Sulfate (Root Killer)**	**Calcium Chloride (Damprid)**	**Hand Sanitizer (with alcohol)**
Goggles	**Pot Holder**	**Spoon**	**Long Match or Lighter**	**Disposable pie tin, cookie sheet, etc**

Procedure:

1. Put your flame plate (Disposable pie tin, cookie sheet, or something the heat from a fire will not damage) on top of a pot holder to protect your work surface.
2. Place a spoonful of one of the chemical salts on your flame plate.
3. Place one squirt of hand sanitizer on this chemical salt.
4. Dim the lights.

5. Using a long match or a lighter carefully light the hand sanitizer on this chemical salt.
6. Observe the color it burns and record it in the data table below.
   Safety note: Do not use or add more hand sanitizer while the flame is still burning!
7. When the flame completely burns out leave the salt on your flame plate.
8. Repeat steps 1-7 with another chemical salt, placing it in a separate pile from the last chemical salt you tested. Do this until you have flame tested each of your chemical salts.
9. To see all the flaming colors together, add one squirt of hand sanitizer to each of your chemical salts (you do not need to add more) and follow steps 2-5.
10. You can repeat this test as many times as you want. Always make sure the flame is completely burned out before adding more hand sanitizer.

Data Table:

Chemical Salt	Color
Sodium Chloride (Table Salt)	
Potassium Chloride (Salt Substitute)	
Copper Sulfate (Root Killer)	
Calcium Chloride (Damp Rid)	

Questions:

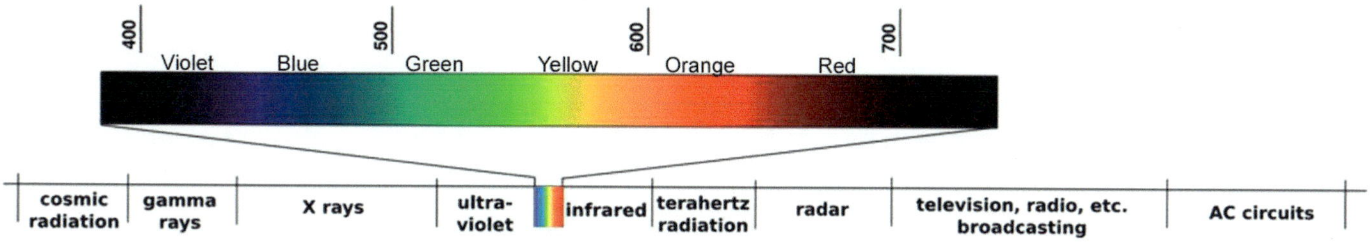

1. Looking at the visible light spectrum, the wavelengths on the right side have a higher frequency and more energy while those on the left side have a lower frequency and lower energy. Which chemical salt's electrons are releasing the most energy as they fall back down to their normal level?

2. A forensic scientist does a flame test on a chemical found at a crime scene. What might they conclude if the light emitted by the substance is green?

3. What is the normal, most stable state for an atom's electrons to be in?

4. What is an electron's state when energy is added and the electron jumps up a level?

5. Explain how the colors in a flame test are produced.

6. Why is a flame test an important lab tool?

Build an Atom

Name:

Go to: tinyurl.com/2vp4nddp

Click on the link above and choose start to explore different atoms and their elements. Fill out this data table as you complete the online lab.

Element	Weight	Material	Color	Symbol	Atomic #	Proton #	Neutron #	Electron #
Helium								
Carbon								
Gold								
Uranium								

Use what you learned in the online lab to build these atoms or to fill in the periodic square:

__ Li
Lithium
____

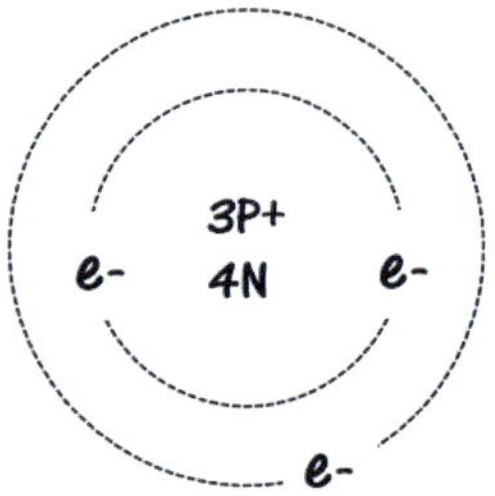

5 B
Boron
11

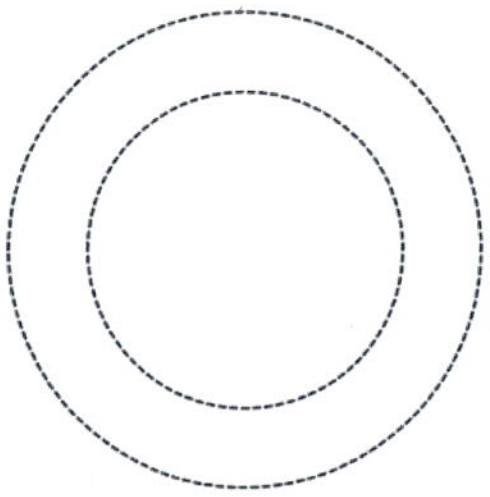

__ O
Oxygen
____

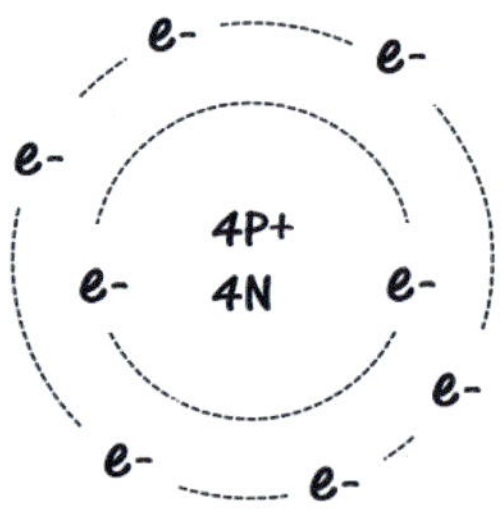

1 H
Hydrogen
1

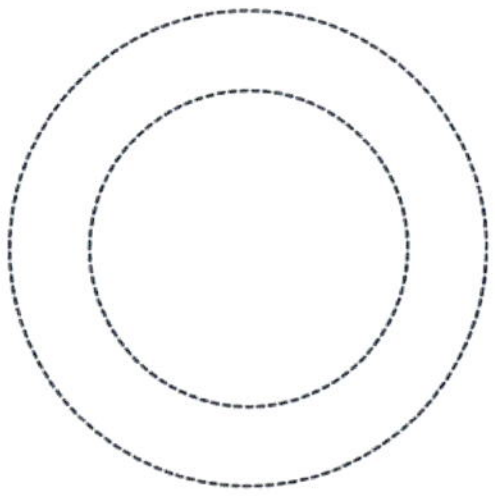

9 F
Fluorine
19

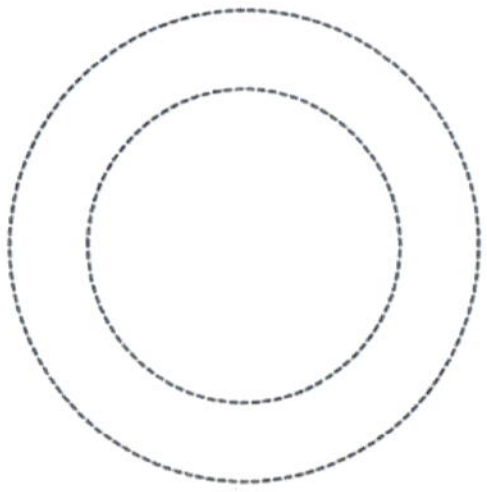

__ N
Neon
____

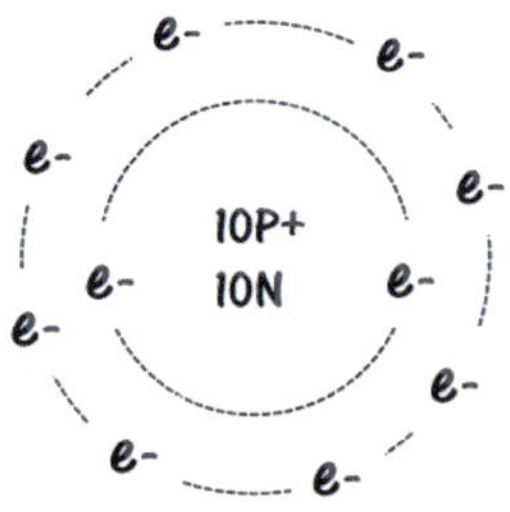

__ Be
Beryllium
____

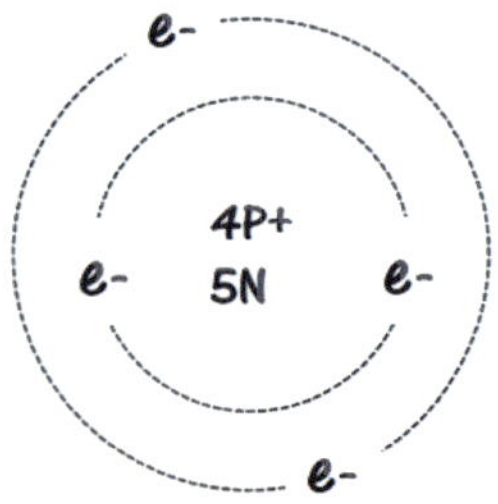

2 He
Helium
4

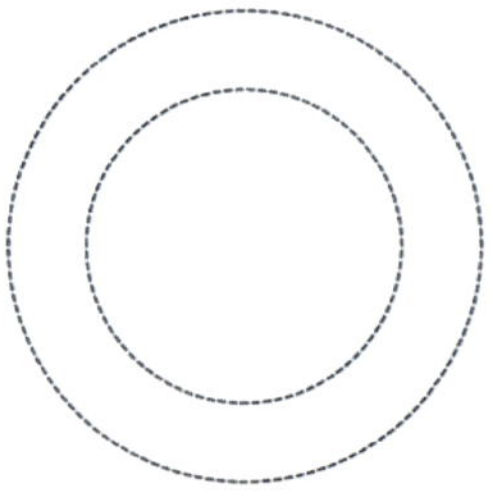

Atomic History Timeline

You are going to create an online timeline to show how the atomic theory has changed over time and to highlight scientists who contributed to the atomic theory. You will use a timeline maker by ReadWriteThink; it is simple and free. You can also opt to create this timeline using any other online timeline tool, or even draw it out on paper.

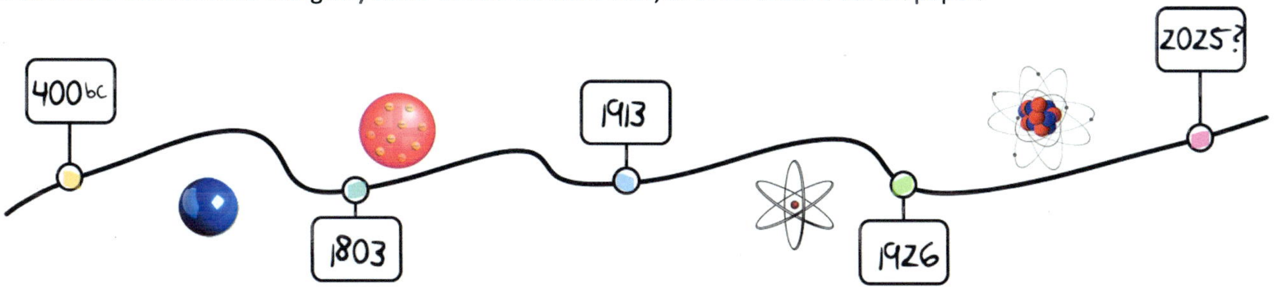

As you learn about the history of the Atomic Theory you see the scientific process in action. As scientists solidify their theories they share them with the scientific community. Other scientists test and build upon these theories using new technologies and ideas. This can lead to a new theory. In science, we are always trying to further our knowledge by experimenting, learning and sharing.

Watch this short video on the history of the atom: tinyurl.com/mpnk5k3

Listed below are the six models of the atom and seven scientists who greatly contributed to our understanding of the atomic theory.

- Democritus
- John Dalton
- J.J. Thomson
- Ernest Rutherford
- Niels Bohr
- De Broglie / Schrodinger

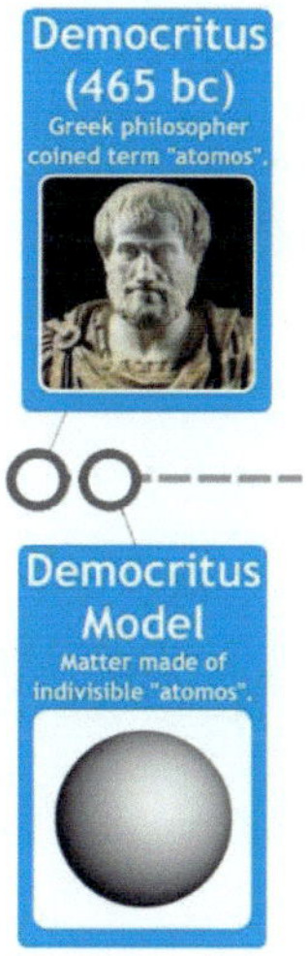

For each of the seven scientists above you must include:

- Under label: their name and either the date of the scientist's birth or of theory.
- Also under label: one sentence description of them, in your own words.
- A picture.
- Place links to any of your sources in the full description box.

For each of the six atomic models you must include:

- Name of model as your label.
- A one sentence description of this model in your own words, also under the label.
- A picture example.
- Credit your sources by placing links to them in your sources in the "full description" box.

Research Tips: When searching online you might get better results if you include the words "atomic theory" in your searches. For example, "atomic theory history" or "Democritus atomic theory." You can find this information by searching online. Some sites we recommend for good information are on our resource page tinyurl.com/2m5fcyfs (password: Matter). When you search make sure that you are using reputable sites. Ask sites and wiki sites are generally poor or unacceptable sources of information for a class project.

To make your timeline go to: tinyurl.com/k477ychm

Click on the button to allow adobe flash. If you do not have this program, you will need to download adobe flash.

- Click on the timeline to add a scientist or atomic model.
- Arrange your timeline so that the scientists and models are in the correct order.
- When you are finished click save and follow the instructions to save the file to your computer.

Picture tip and tutorial: When you find a picture online, right click on it and choose "save as" to save the picture to your computer. You will then have to drag your pictures to where it says to drag and drop them on the timeline. You can watch this video to see how we set up our timeline: tinyurl.com/24hymsxa

The Periodic Table

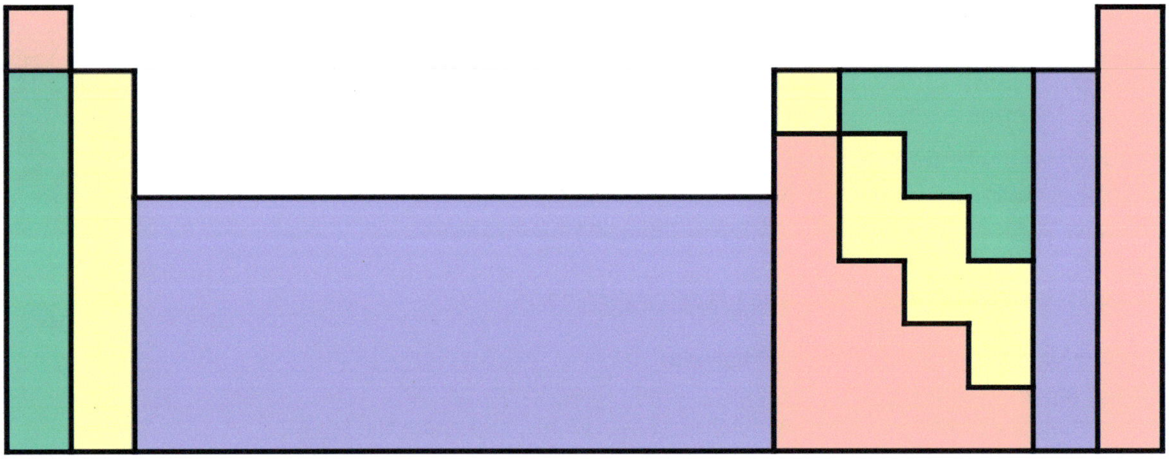

Periodic Table

6 — Atomic Number = # Protons
Protons = # Electrons
(in balanced atom)

C — Symbol

Carbon

12 — Atomic Mass = Protons + Neutrons

Groups / **Periods**

Periods	1	2	3	4	5	6	7	8	9	10	11	12	13	14	15	16	17	18
1	1 H Hydrogen 1																	2 He Helium 4
2	3 Li Lithium 7	4 Be Beryllium 9											5 B Boron 11	6 C Carbon 12	7 N Nitrogen 14	8 O Oxygen 16	9 F Fluorine 19	10 Ne Neon 20
3	11 Na Sodium 23	12 Mg Magnesium 24											13 Al Aluminum 27	14 Si Silicon 28	15 P Phosphorus 31	16 S Sulfur 32	17 Cl Chlorine 35	18 Ar Argon 40
4	19 K Potassium 39	20 Ca Calcium 40	21 Sc Scandium 45	22 Ti Titanium 48	23 V Vanadium 51	24 Cr Chromium 52	25 Mn Manganese 55	26 Fe Iron 56	27 Co Cobalt 59	28 Ni Nickel 59	29 Cu Copper 64	30 Zn Zinc 65	31 Ga Gallium 70	32 Ge Germanium 73	33 As Arsenic 75	34 Se Selenium 79	35 Br Bromine 80	36 Kr Krypton 84
5	37 Rb Rubidium 85	38 Sr Strontium 88	39 Y Yttrium 89	40 Zr Zirconium 91	41 Nb Niobium 93	42 Mo Molybdenur 96	43 Tc Technetium 98	44 Ru Ruthenium 101	45 Rh Rhodium 103	46 Pd Palladium 106	47 Ag Silver 108	48 Cd Cadmium 112	49 In Indium 115	50 Sn Tin 119	51 Sb Antimony 122	52 Te Tellurium 128	53 I Iodine 127	54 Xe Xenon 131
6	55 Cs Caesium 133	56 Ba Barium 137		72 Hf Hafnium 178	73 Ta Tantalum 181	74 W Tungsten 184	75 Re Rhenium 186	76 Os Osmium 190	77 Ir Iridium 192	78 Pt Platinum 195	79 Au Gold 197	80 Hg Mercury 201	81 Tl Thallium 204	82 Pb Lead 207	83 Bi Bismuth 209	84 Po Polonium 209	85 At Astatine 210	86 Rn Radon 222
7	87 Fr Francium 223	88 Ra Radium 226		104 Rf Rutherfordium 267	105 Db Dubnium 268	106 Sg Seaborgium 269	107 Bh Bohrium 270	108 Hs Hassium 277	109 Mt Meitnerium 278	110 Ds Darmstadtium 281	111 Rg Roentgenium 282	112 Cn Copernicium 285	113 Nh Nihonium 286	114 Fl Flerovium 289	115 Mc Mozcovium 290	116 Lv Livermorium 293	117 Ts Tennessine 294	118 Og Oganesson 294

57 La Lanthanum 139	58 Ce Cerium 140	59 Pr Praseodymium 141	60 Nd Neodymium 144	61 Pm Promethium 145	62 Sm Samarium 150	63 Eu Europium 152	64 Gd Gadolinium 157	65 Tb Terbium 159	66 Dy Dysprosium 163	67 Ho Holmium 165	68 Er Erbium 167	69 Tm Thulium 169	70 Yb Ytterbium 173	71 Lu Lutetium 175
89 Ac Actinium 227	90 Th Thorium 232	91 Pa Protactinium 231	92 U Uranium 238	93 Np Neptunium 237	94 Pu Plutonium 244	95 Am Americium 243	96 Cm Curium 247	97 Bk Berkelium 247	98 Cf Californium 251	99 Es Einsteinium 252	100 Fm Fermium 257	101 Md Mendelevium 258	102 No Nobelium 259	103 Lr Lawrencium 266

Printable available on Online Resource page

A Special Table

Name:

Watch: tinyurl.com/yc256kc9

1. A substance made of only one kind of atom is an: ________________________

2. The periodic table is arranged from the __________________ to the __________________ atoms.

Label the parts of the element square. →

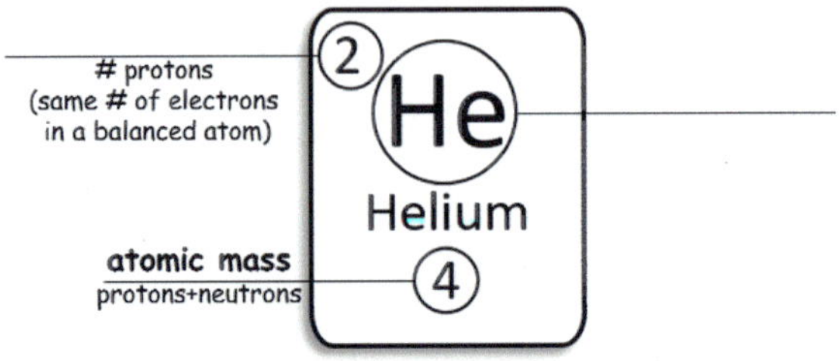

Number the groups and periods (rows). Write the family name's in their sections of the periodic table (pause the video to give yourself time).

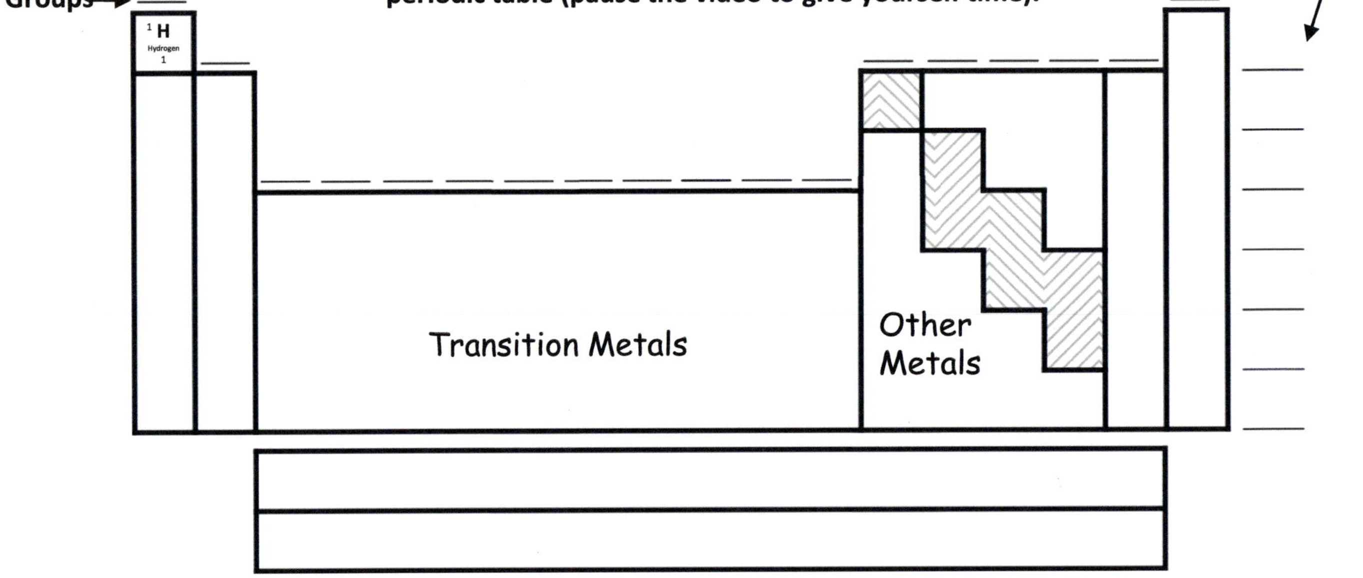

3. What does the diagonal band of elements split the periodic table into?

4. What are 4 properties that metals have?

5. Lightly color the metals gray. Label and color the nonmetals yellow and the semimetals blue.

6. The universe is mostly what element? What makes this element special in other ways?

<u>Watch "10 strangest elements"</u>: tinyurl.com/3ab9dcjh

Pick three of these super cool elements to take notes on.

Name	Description	Cool Fact #1	Cool Fact #2	Use

Periodic Scavenger Hunt

Name:

Go to: tinyurl.com/oun492s

Click on the element squares of this site to label this periodic table and to answer the questions below. You will notice that some families such as the semi-metals in your notes are called metalloids.

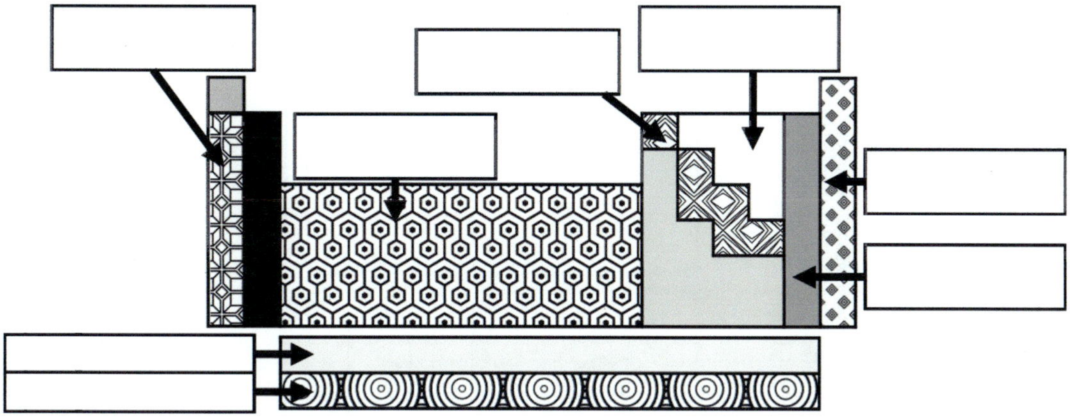

1. Click on Lithium. What are three examples of how we use lithium?
   - What color is lithium in its purest form? Is it a Solid, Liquid or Gas? (**hint:** click interactive for the symbol key.)
   - Lithium is in the Alkali Metal family. What are 3 facts about Alkali Metals? Include what Lithium catches fire with.
2. Which elements are Alkali metals? List their symbols: _____ _____ _____ _____ _____ _____
3. What are 3 characteristics of Alkali Earth Metals? How reactive are they?
4. Describe Calcium. What color is it in its purest form?
5. If you wanted to conduct electricity well would you use a Transition Metal or an Alkali Earth Metal?
6. List three examples of a transition metal. From their description do you think they are reactive?
7. Do poor metals melt more easily than transition metals? Between Gallium and Zinc which do you think will have a lower melting point?
8. What is special about Metalloids? Do they conduct electricity?
9. Which elements are Metalloids? List their symbols: _____ _____ _____ _____ _____ _____
10. What does the term "Halogen" mean?
11. What is something Noble Gases never do? Why is this?
12. Which is your favorite Noble Gas? Why?
13. What is unique about Hydrogen's periodic family (group)?
14. How much of the universe is made of Hydrogen atoms?
15. Which elements are gases at room temperature? ____ ____ ____ ____ ____ ____ ____ ____ ____ ____ ____

Check out some of these awesome periodic table videos by the university of Nottingham. Here you can see real elements and how they react and even explode! periodicvideos.com

Atom's Family Electrons

Name:

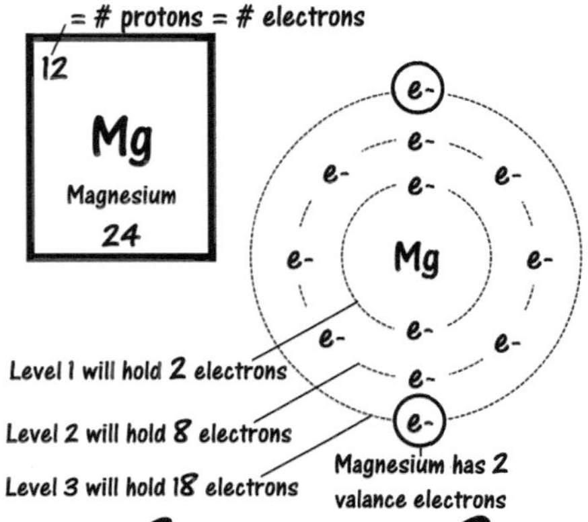

Use your periodic table to find the atomic # and then to draw in the electrons for each of the elements below. What do you notice about the number of valance electrons in each periodic group?

Valance electrons are the electrons in the outside energy level or shell of an atom. Each energy level can only hold a certain number of electrons. Atoms really want to have full energy levels, so they give or take electrons from other atoms by bonding with them. The **most reactive elements only need to get rid of or gain one more electron** to have a full energy level (groups 1 and 7). Atoms with full energy levels do not bond with other atoms and are not reactive (**ex: group 8: noble gases**).
As you move down to the 4th and 5th periods of the periodic table the atoms will have more levels. We are focusing on the first 3 periods because they are easier to draw.

Groups 1	2	3	4	5	6	7	8
H							He
Li	Be	B	C	N	O	F	Ne
Na	Mg	Al	Si	P	S	Cl	Ar

Each element box: Name ____ Atomic # ____ Valance e- ____

Periodic Table Battleship

You are going to play battleship using your periodic table! Try to hit your opponent's battle ships by calling out the elements or the groups and periods. The first one to sink all the ships wins!

Play this game like traditional battleship in groups of two. Each player will have their folder open so that their opponent cannot see their periodic tables.

target grid

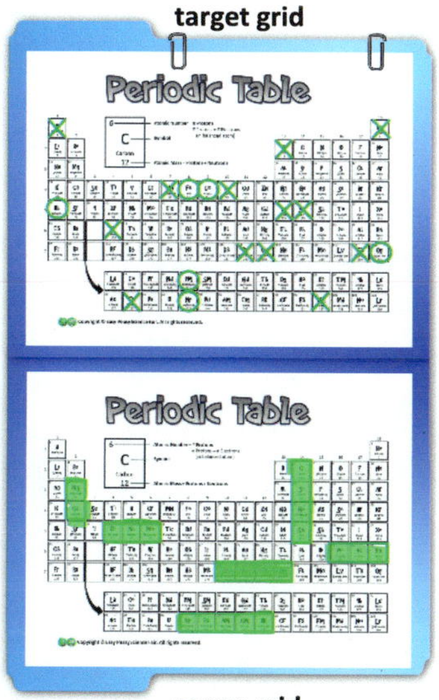

ocean grid

1. Use your marker to outline your element ships on the "ocean grid". You can place your ship either vertical or horizontal, not diagonal.
   - 1 Aircraft Carrier = 5 elements
   - 2 Battleships = 4 elements
   - 2 Submarines = 3 elements
   - 1 PT boat = 2 elements
2. The first player takes a shot calling the period and group for the element of their choice. The other player states the name of the element called, to confirm, and then says "hit" or "miss."
3. Mark hits on the "target" grid by circling the elements and place an **X** on the elements for misses. These are your guesses.
4. As your opponent calls their shots, mark their hits and misses on your ocean grid to help you keep track of what they are doing.
5. If all the elements of a ship have been guessed, it is sunk. Let your opponent know when they sink a ship.
6. The first player to sink all opposing ships wins the game!

Try playing again. This time calling out the atomic symbols, atomic numbers or masses. You can also play by calling out the number of protons or neutrons, for a challenge.

Student instructions

Practice your periodic table skills as you play this fun game. Once this game is over you should be able to locate elements on the periodic table by their atomic symbol and their position on the table.

Play time: 20-30 minutes
Materials for 2 Players:
- 2 file folders
- 2 dry erase markers
- 4 large paper clips or tape
- 4 copies of the periodic table, laminated
  (You can also use page protectors or in a pinch you can use a gallon Ziploc bag or cover them with saranwrap.)

Vocabulary:

Atomic Symbol	Protons
Atomic Mass	Electrons
Period	Neutron
Groups	

Physical Properties & Changes

Physical Property Vocab

Cut and paste the definitions and pictures that go with these physical properties. **Name:**

You can use the internet or resource books to find the definitions.

1	Physical Properties		
2	Viscosity		
3	Freezing Point		
4	State of Matter		
5	Luster		
6	Ductility		
7	Volume		
8	Buoyancy		
9	Conductor		
10	Magnetic		
11	Mixture		
12	Density		
13	Solubility		
14	Malleability		

Cut and Paste Pictures and Definitions

Use the internet or reference books to match up these definitons and pictures with the terms on the vocabulary page.

The ability of an object to float in water or air.	Allows electricity to flow through it easily.
A measure of how compact an object is.	The ability to be hammered or shaped under pressure without breaking.
solids, liquids, and gases	When 2 or more substances combine and a chemical change hasn't occurred.
The temperature at which the liquid freezes. Water freezes at 0° Celsius.	Attraction to a magnet.
Capable of being shaped or bent.	How much space an object takes up.
The degree of shininess of an object. Ex: bright or dull	Can be observed without chemically changing the matter.
The ability to be dissolved easily.	The resistance of a liquid to flow.

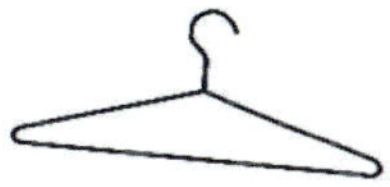

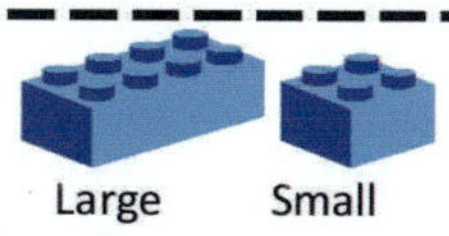

Physical Property

Ooey Gooey Oobleck Lab

Name:

We are going to make and investigate the physical properties of Oobleck to see if it is a solid, liquid or a gas.

Goggles on! Tie hair back.
Do not eat Oobleck or leave it on carpets, furniture or other fabrics. To keep Oobleck in playing condition store it in a Ziploc bag in the refrigerator.

Background Research - List three properties of solids, liquids and gases:

Solids	Liquids	Gases

Materials:

Ziploc bag	Borax	Food Coloring	White Elmer's Glue
Warm Water	¼ and 1 Tablespoon	Spoon	1 Cup

Prepare Liquids:

1. Add 2 Tablespoons of warm water with 2 Tablespoons of Elmer's glue in your cup and stir until completely smooth. This is your white liquid.
2. Clean the Tablespoon and add 4 Tablespoons warm water and ¼ teaspoon of Borax to your bag. Seal bag and mix until the Borax is dissolved. This is your clear liquid.

Mix a White Liquid, a Clear Liquid and a Colored Liquid. What State will Oobleck be?

1. Open the plastic bag.
2. Choose and add 4 drops of the liquid food coloring.
3. Carefully pour the white liquid into the plastic bag.
4. Close the bag and knead the mixture well for 3 minutes.
5. Once 3 minutes have passed, take the Oobleck out of the bag and complete the investigation of its properties.

Investigation of Oobleck's Physical Properties:

1. Use your senses and measuring tools to identify the various physical properties of your Oobleck.
2. Record them in the table. Remember to use phrases such as "it smells like" and "it feels like."

Property	Observation	State of Matter
Example: Color	It looks green.	S L G
Length it Stretches		S L G
Bounce		S L G
Color		S L G
Smell		S L G
Texture		S L G
Shape		S L G
Ductility		S L G
Imprint Writing		S L G
Tear		S L G
Compress		S L G

Take Container Shape		**S L G**
Malleability		**S L G**
Viscosity high/low		**S L G**
Squeeze with fist		**S L G**
Keep shape when held for a minute		**S L G**

Conclusion:

From the data collected Oobleck appears to be a ________________. (Solid, Liquid or Gas)

Questions:

1. What type of properties were identified in today's investigation? ______________________________
2. Put a **star** by the properties identified through quantitative observations and an **X** by properties identified through qualitative observations.
3. If you measured the mass of your Oobleck, what unit would you use? ___________ Would this observation be quantitative or qualitative? __________________________
4. What is Ductility? ____
5. If you put your Oobleck in the freezer overnight, would its ductility change? ________________________
6. When two or more elements are joined together with chemical bonds, they form a compound. On the other hand, when two or more elements are mixed together without chemical bonds, they form a mixture. Based on this information, is your Oobleck an element, a mixture, or a compound? __________________________
7. In the space provided, write a descriptive paragraph about your Oobleck using all the properties identified in this investigation.

Density Challenge Online Lab

Name:

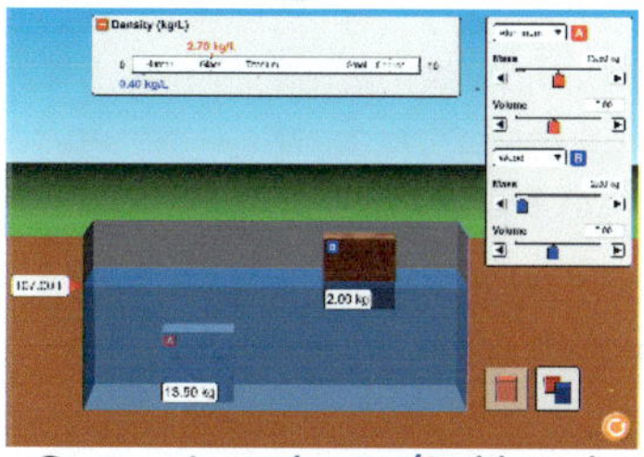

Go to: tinyurl.com/3ukbznzk

Density is a physical property of matter. It is how close the molecules of a substance are. Another way of saying this is how much mass an object has in a given space. What are two objects in your house that you think will sink in water and two you think will float?

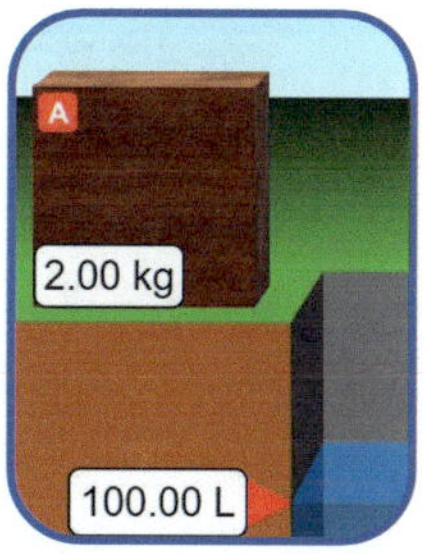

1. There is a number on the block of wood. This number represents the wood's mass. **Mass** is the amount of matter, or "stuff" in an object. What is the mass of the block of wood? Put a "kg" (kilograms) after your number; this is the measurement unit used for mass.
2. Take your block of wood out of the water the water level next to the arrow should now read 100.00 L. Pull your block of wood all the way down into the tank of water. What does the water volume on the side read now? Write this below under volume with object.

____ L	-	100 L	=	________
volume with object		**starting volume**		**volume of object**

3. Subtract the starting volume of the water from the volume of the water with the object. This will give you the actual volume of the block of wood. **Volume**, is the amount of space an object takes up. Put L (liter) after your number. This or cm³ is the measurement unit for volume. 1 cm^3 = 1 mL
4. Use the pull-down arrow in the upper right to select other objects to measure. Fill in the data table as you go.

Material	Result (sink or float?)	Mass (kg)	Volume (L)	Density (kg/L)	Density Less than...
Wood	float	2.00 kg	5 L	.40 kg/L	a human
Styrofoam					
Ice					
Brick					
Aluminum					

Look at the data in your table.

4. Can you use just the volume of an object to predict whether it will sink or float?
5. Choose the two-block option on the bottom right of your screen. Make **item A** wood and **item B** Aluminum. Adjust the mass of each so that they are the same. **Wood:** ________ (mass) ________ (**sink/float**) **Aluminum:** ________ (mass) ________ (**sink/float**)
6. Can you use just the mass to predict whether an object will sink or float?
7. Can you use mass **and** volume to predict whether an object will sink or float? Explain.
8. In your data table, circle or put a star next to any item that has a density less than 1 kg/L. Do these items sink or float?
9. **Density** is the amount of mass in a specific volume. To find the density, divide the mass by the volume. The unit for density is kg/L or kg/cm³. What is the density of a ball that has a mass of 6 kg and a volume of 2 L? ________ **sink/float**
What is the density of a toy that has a mass of 1 kg and a volume of 2 L? ________ **sink/float**
10. The density of water is 1.0 kg/L or 1.0 kg/m³. Use density to make a rule about whether an object will sink or float.

Physical Properties of Matter Project

There are so many cool physical properties and changes of matter to see. Make a presentation to show us all about these physical properties and their changes! Use printed resources or the internet to gather your information for this project. Instead of a quiz this week you will present your project.

You will find instructions for making a Prezi presentation below, but with permission you can make a poster, PowerPoint, or some other project instead. Your presentation needs to include:

- ☐ A title.
- ☐ Your first name.
- ☐ Description of what a physical property of matter is (in your own words).
- ☐ Insert a YouTube video on physical properties or changes. (If you are making a poster, you can just use a picture).
- ☐ Description of 4 ways that physical properties are observed and measured.
- ☐ 5 types of physical properties, with a picture example of each.
- ☐ 3 examples of physical changes in real life, with a picture example of each.
- ☐ Spelling, grammar and creativity will be a part of your grade.

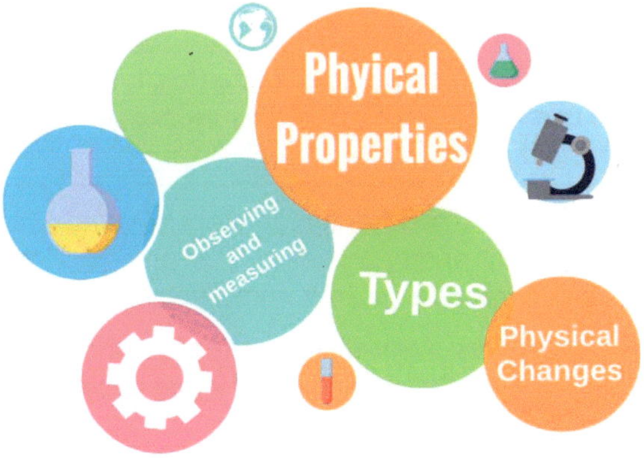

You can find this information by searching online. Some sites we recommend for good information are on our resource page tinyurl.com/2m5fcyfs (password: Matter). When you search make sure that you are using reputable sites. Ask sites and wiki sites are generally poor or unacceptable sources of information for a class project.

Making a Prezi

Prezi is a really cool interactive slideshow that you can make for free. Here are the instructions to get you started!

1. Start by signing up for a **free** "basic" Prezi account: prezi.com/signup/basic
   a. **You will need an email address for this part.** You can use your parent's or your own email address.
2. We recommend watching the "getting started" Prezi video so that you can familiarize yourself with how to Prezi.
3. When you are ready, click on the + sign and choose presentation.
4. Browse through the templates until you find the one you like. Click on it then select "use this template."
5. Name your presentation and choose to let anyone on the web see your presentation.
6. Take a few minutes to explore your Prezi and to figure out where you want your topics to be placed.
   a. To zoom in, click on the slides to the left.
   b. Return to your big picture by clicking [Go back] or "go back to parent".
   c. Add new slides by clicking [+ Topic] or by clicking the topic arrow down and choosing [Zoom area]
   d. Change the order of your slides by clicking on a slide and dragging it up and down the left-hand column.
7. To write on a slide click on the slide and then click where you see text. You can add new text by clicking insert and choosing "insert text."
8. To insert a picture, click on insert and choose "insert picture." You will then choose a picture on your computer. Resize it by clicking on a corner and pulling it in or out. Then place the picture where you want it to go. You can use this tutorial on how to save images onto your computer: tinyurl.com/ycxbm2lr
9. To insert a video, first find a video on youtube that you like and copy the link (Ctrl+C or right click and copy). Then click on "insert," choose "insert video," and paste your link in the box (Ctrl+V or right click and paste).

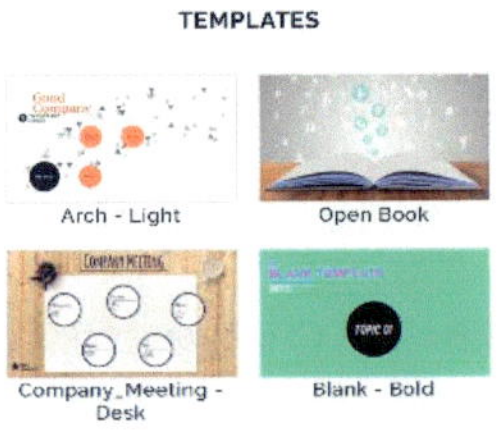

Chemical Properties & Changes

Cool Facts and Chemical Properties

Name:

Properties of Matter are the characteristics and behaviors we use to describe matter. There are two types of properties: **Chemical** and **Physical**.

Each substance has specific **properties** which help us to identify it and use it. For example Red food coloring has a red color (physical property) but it loses its color as it reacts with bleach (chemical property).

Chemical changes result in an entirely new substance. These changes are usually more obvious than a physical change. You cannot reverse a chemical change. Examples of a chemical change would be rusting and burning.

Clues that a chemical change have happened are:

- Unexpected color change
- Change in smell
- Fire is produced
- Gas bubbles appear
- Electricity or energy is made
- Light is produced
- Solid forms from liquids
- Unexpected temperature change

Write a "C" if the cool fact illustrates a chemical property or change and/or a "P" if it illustrates physical instead. Explain your choice in the line below each fact.

_____ 1. Exploding fireworks used to be only orange and white. In the Middle Ages new colors were made by burning different salts. The hardest color to create is blue.

____

_____ 2. Liquid mercury is so dense that you can float a coin in it.

____

_____ 3. A piece of toast was burned in the shape of the "Virgin Mary" and then sold for $28,000!

____

_____ 4. I picked up a metal chunk of the element Gallium and it melted from the heat of my hand.

____

_____ 5. You can make steel and iron rust in just 10 minutes. Soak it in vinegar for five minutes. Remove it, then cover the metal with salt and hydrogen peroxide for five minutes.

____

_____ 6. One inch of liquid rain is equal to 10 inches of frozen snow!

____

_____ 7. You can cut the metal potassium with a knife but when you drop it in water it will burst into flames!

____

_____ 8. With the right amount of pressure and heat, the graphite in your pencil will turn into a diamond!

____

_____ 9. Almost every day of his life, Albert Einstein fried up and ate two eggs.

____

_____ 10. A team of scientists in Georgia painted a picture of the Mona Lisa that is is a third the width of a human hair. It is the world's smallest painting!

____

_____ 11. If you chew a piece of a cracker in your mouth for a minute it will start to taste sweet. This is because the amylase in your saliva breaks down the cracker starch into sugars.

____

_____ 12. A handful of salt poured and mixed into water the makes the water level go down instead of up.

____

Crazy Elephant Toothpaste Lab

Name:

We are going to create a cool and surprising chemical reaction.

Goggles on! Tie hair back.
Hydrogen peroxide can irritate skin and eyes and bleach clothes, it is important to wear goggles the entire time. This lab can be messy.

Background Info:

When a **chemical reaction** happens, energy is either released or absorbed. Energy is usually released as heat energy and the temperature will go up and feel warm or even extremely hot. This is an **exothermic reaction**. The opposite of an exothermic reaction is an endothermic reaction. In an **endothermic reaction,** energy is absorbed, and the substance and surroundings will feel colder.

During a chemical reaction substances called **reactants** change into new substances called **products.** These new products have their own properties and might change volume, become a gas or turn a different color.

Chemists use **catalysts** in reactions. A catalyst will boost a reaction, making it happen more quickly. This happens because a catalyst either allows a reaction to happen with less energy, or it provides the energy needed for that reaction.

Materials:

40 Volume Hydrogen Peroxide (12%)	**Rapid Rise or Active Yeast**	**Food Coloring**	**Dawn Dish Soap**	**20 oz Soda Bottle, empty**
Casserole Dish or Tray	**Tablespoon and ½ Cup**	**Cup**	**Very Warm Water**	**Dawn Dish Soap**

1. Place your empty soda or water bottle in the center of your casserole dish or tray.
2. Pour a ½ cup of the hydrogen peroxide into the bottle.
3. Add 4-5 drops of food coloring into the bottle and a Tablespoon of the dish soap. Gently swirl the bottle to mix your ingredients.
4. Add 3 Tablespoons of very warm water to your cup. pour a packet of yeast into the cup and stir for 30 seconds (sing the alphabet) until there are no clumps in your solution.
5. Pour your yeast mixture into the soda bottle and watch what happens!
6. Feel the foam and the bottle. Would you describe the temperature as warm or cool?
7. Write down your observations. Measure the volume of the foam with a beaker or measuring cup, if you can, and include this in your observations.

Questions:

1. After the reaction, was the temperature of the bottle and the foam warm or cool? Would you say this was an endothermic or an exothermic reaction? Why?

2. The gas produced from the reactants in this chemical reaction is a _____________? Which of the reactants would you say was responsible for this gas?

3. What is a catalyst? What do you think the catalyst was for this reaction? Why?

4. In this experiment hydrogen peroxide decomposes to form water and oxygen. Balance the chemical equation, as you learned how to do in the "Chemical Chase and Changes" online lab:

$$2H_2O_2 \rightarrow \_\_\_\_ H_2O + \_\_\_\_ O_2$$

5. What is one variable you think could be changed to create more foam?
If you tested this, the variable you change would be your _____________ variable. What would be your dependent variable? **Challenge: Run an experiment to test this.**

Chemical Chase and Changes

Name:

Chemical and Physical Change Examples

Go to: tinyurl.com/zfvtxbuk

Define each type of change:

Physical	Chemical
Causes of Physical Change: • • •	Clues of Chemical Change: • • •

1. Can a **chemical change** be reversed?

Complete the data table:

Example	Type of Change	Explanation
2. Broken Cracker		
3. Burning Log		
4. Rust		

Complete Matching
Go to: tinyurl.com/nm39avmh

Example	Type of Change	Explanation
5. Boiling Water		
6. Sour Milk		
7. Cooking an Egg		
8. Cutting an Apple		
9. Fireworks		
10. Folding Laundry		

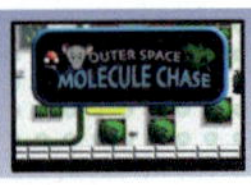

Molecule Chase
Go to: tinyurl.com/3ckevzct

Play the game to build molecules with the atoms you find. Identify the two molecules below.

11. ____________

12. ____________

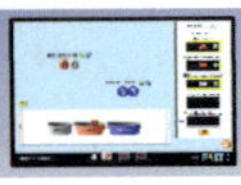

Build a Molecule
Go to: tinyurl.com/yp6rrjcj

Drag the atoms up to create the required molecule. Click between the atoms to create the bonds. Choose "Single".

Click 3D to see the ball and stick version of this molecule

13. ______________________ **name of molecule**

H_2O **chemical formula** ________ **# of elements** ________ **# of atoms**

14. ______________________ **name of molecule**

________ **chemical formula** ________ **# of elements** ________ **# of atoms**

15. ______________________ **name of molecule**

________ **chemical formula** ________ **# of elements** ________ **# of atoms**

16. ______________________ **name of molecule**

________ **chemical formula** ________ **# of elements** ________ **# of atoms**

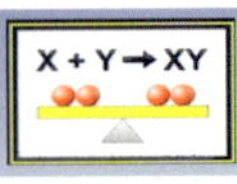

Balancing Chemical Equations
Go to: tinyurl.com/4janwzdh

Enter the numbers for each molecule that balanced your Equation. Hint, there will be the same number of each atom on either side. Choose "Introduction".

17. ______ N_2 + ______ H_2 = ______ NH_3
Nitrogen + Hydrogen = Ammonia

18. ______ H_2O = ______ H_2 + ______ O_2
Water = Hydrogen + Oxygen

19. ______ CH_4 + ______ O_2 = ______ CO_2 + ______ H_2O
Methane + Oxygen = Carbon Dioxide + Water

Chemical Properties of Matter Project

Part 2

There are so many cool chemical properties and changes of matter to see. Show us about chemical properties and their changes by adding them to the physical properties project you did in the last unit. Instead of a quiz this week, you will present your project.

You will find instructions for adding onto your Prezi presentation below. With permission you can add onto some other project, such as a poster or PowerPoint. Your presentation needs to include:

- ☐ "Properties of Matter" title.
- ☐ Description of what a chemical property of matter is (in your own words).
- ☐ Insert a YouTube video on chemical properties or changes. (If you are making a poster, you can just use a picture).
- ☐ Description of ways that chemical properties are observed and measured.
- ☐ 3 types of chemical properties, with a picture example of each.

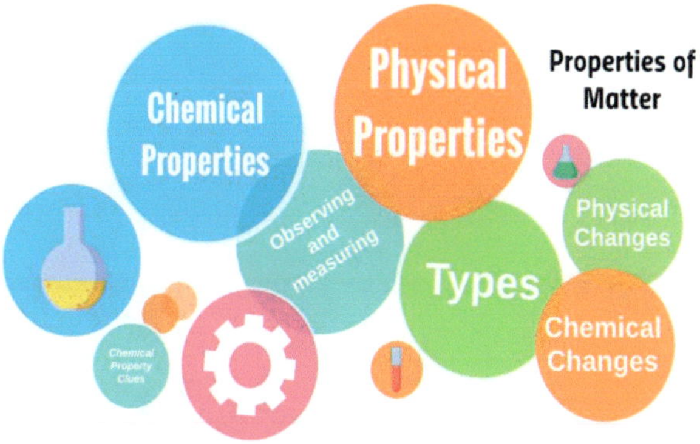

- ☐ 3 examples of chemical changes in real life, with a picture example of each.
- ☐ 8 clues of a chemical change, with a picture example of each
- ☐ State somewhere, what you think makes a chemical change different from a physical change.
- ☐ Spelling, grammar and creativity will be a part of your grade.

You can find this information by searching online. Some sites we recommend for good information are on the resource page tinyurl.com/2m5fcyfs (password: Matter). When you search make sure that you are using reputable sites. Ask sites and wiki sites are generally poor or unacceptable sources of information for a class project.

Making a Prezi -Part 2

Try some of these new tips and tricks to make your Prezi even better!

1. Click the down arrow on [+ Topic ▾] and then choose [Planet topic] to add a big topic slide to your presentation.
2. Add new text by clicking insert and choosing "insert text."
3. 

4. Try rotating your content: 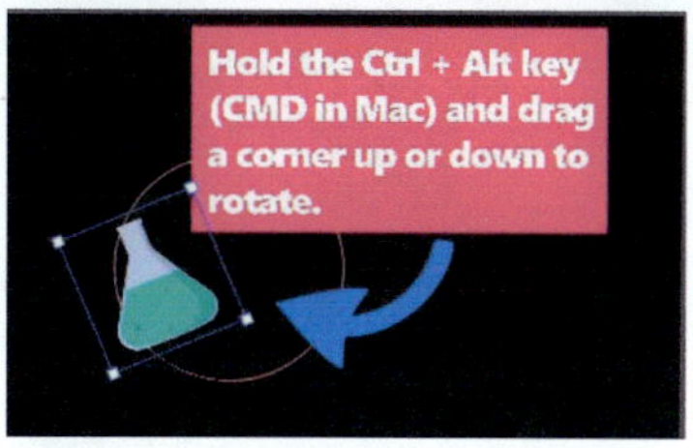

Click on any object and add an animation to it.

a. You can choose "Fade in" or fade out to make objects and text appear and disappear during your presentation.
b. You can also choose a "Zoom" to zoom into or out of an area during your presentation.
Tip: to adjust the zoom area just change the size of the rectangle that appears when you choose to zoom.

Acids & Bases

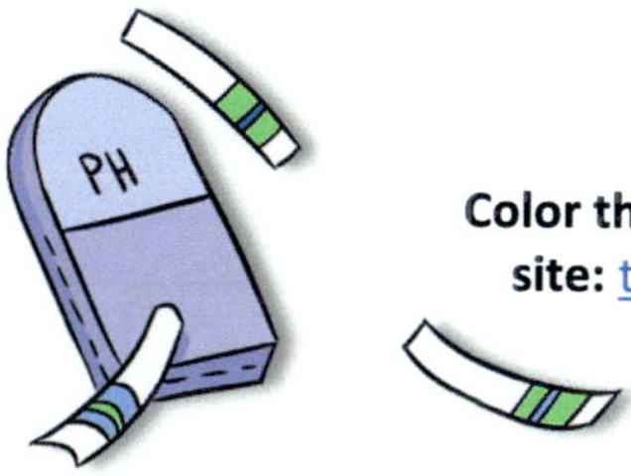

Acids & Bases

Name:

Color the pH scale below and write an example for each pH level using this site: tinyurl.com/yb38mxwj. Label the Base, Acid and Neutral Sections.

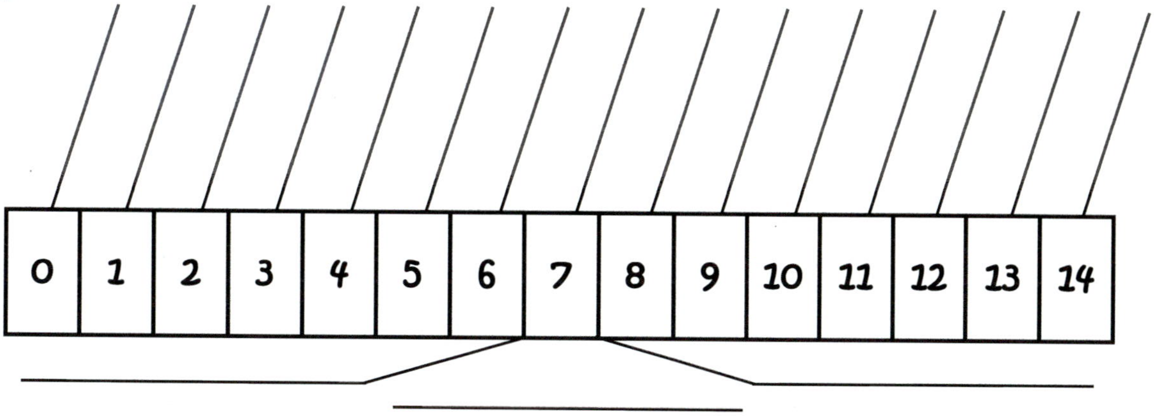

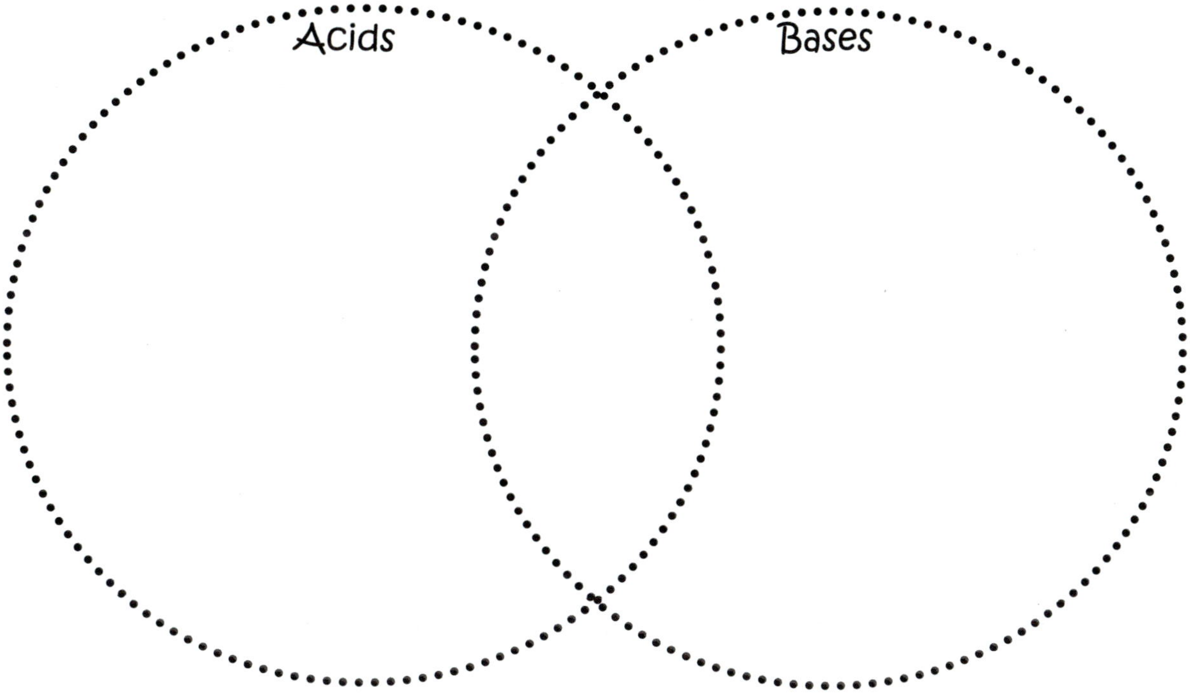

Cut out the properties of acids and bases below and place them where you think they belong on the Venn Diagram. Then read this site: tinyurl.com/5cdsbksz to check your work. When all properties are in the correct spot, paste them in place.

pH less than 7

pH more than 7

Reacts with Oils + Fats

Produces Salt when Neutralized

Sour Taste

Feels Soapy

Fruits + Coffee

Can Blister and Corrode

Bee Sting

Neutralizes Acids

Neutralizes Bases

Turns Litmus Blue

Turns Litmus Red

Cleaning Products

ACID ATTACK LAB

Name:

We are going to make our own indicator and test the pH of difference household substances.

Goggles and aprons on! Tie hair back.
Do not drink or eat any of your lab ingredients. Red cabbage juice has a smell and can stain, so be careful with your clothes.

Question: Which substance is the most acidic and will best eat away a piece of chalk or an egg shell?

Background Info:

Acids and bases are all around us. They are in our foods, our pools and even help our bodies digest food. Each has their own properties. For example, acids taste sour and often react with metals while bases taste bitter and have a slippery feel. Strong acids and bases can both be very dangerous and will burn your skin. We measure how acidic or basic a chemical is using a pH scale.

The strength of the pH scale is determined by the kind of ions that are released. Acids release **hydrogen ions (H+)**. The higher the concentration of hydrogen ions (H+) the more acidic something is and the lower its number on the pH scale. Bases release **hydroxide ions (OH-)**. The higher the concentration of hydroxide ions (OH-) the more basic something is and the higher its number on the pH scale.

Different indicators can be used to show how acidic or basic something is. In this lab, you will use a cabbage juice indicator which will turn different colors to show you how acidic or basic something is. **Flavin** is the chemical in cabbage juice that changes colors in acids and bases.

Cabbage juice pH scale:

Acid	Acid	Acid	Acid-Neutral	Base	Base	Base	Base
0-1	2-3	4-5	6-7	8	9-10	11-13	14
Red	Light Pink	Dark Pink	Purple	Blue	Blue-Green	Green-Yellow	Yellow

Materials:

¼ Purple Cabbage	**Warm Distilled Water**	**Blender**	**Mesh Strainer**	**9 Clear Cups, Ice Cube Tray or White Egg Carton**
2 Spoons	**Baking Soda**	**Clear Shampoo**	**Cream of Tartar**	**Ammonia**
White Vinegar	**Sugar**	**Salt**	**Lemon Juice**	**Egg or Chalk Stick**

Prepare Cabbage Juice Indicator:

1. Put ¼ a head of cabbage in your blender and add 4 cups of very warm water. Blend for 1 minute.
2. Pour cabbage juice through the mesh strainer into a larger container.
3. If you want to store the cabbage juice, stir in 2 Tablespoons of salt and refrigerate. Use within 24 hours.

Testing pH of Chemicals:

1. Place 2 spoonfuls of your cabbage juice indicator into one of your cups or ice cube wells.
2. Add one spoonful of baking soda and mix. What color did it turn? Record this in your data table.
3. Test the rest of your chemicals and water (which is a chemical too) the same way. Record all of your data. Do not test the egg or chalk stick, we will use this later in the lab.

Chemical	Cabbage Juice Color After	Conclusion: Acid, Base, Neutral	Ph #

4. Sort your results into the following categories. Underline what you believe to be your strongest acid and your strongest base.

Strong Acids	Weak Acids	Neutral	Weak Bases	Strong Bases

Acid Attack:

5. Place your egg or stick of chalk in a cup. Eggshells and chalk are made of **calcium carbonate**.
6. Pick the strongest acid you tested and pour it into the cup so that the egg is fully covered or so that half of the stick of chalk is in the strong acid.
7. Write down your observations after 10 minutes. Keep your egg or chalk in the acid for 2-3 days and then write down your final observations.

Observations after 10 Minutes	Observations after 2-3 Days

Neutralization: Is the color change from your cabbage juice reversible?

8. Add one of your colored basic chemicals (avoid baking soda - it can get messy) to a colored acid. See if you can get the color back to purple. Were you able to do it? ________ Yes ________ No

Questions:

1. Which of the chemicals you tested has the highest concentration of hydrogen ions (H+)?

2. What type of substance changes color when exposed to an acid or a base?

3. Milk has a pH of 6.8. Is it slightly acidic or slightly basic?

4. The burning of coal and fossil fuels release gases in the air that make our rain slightly acidic. This is called acid rain. What might acid rain do to buildings and statues made of limestone and marble which contain calcium carbonate, just like egg shells and chalk?

Alien Juice Bar

Name:

tinyurl.com/5x5hknwj

Click on the link above. You will need Adobe Flash Player for this GEMS science activity. If you have trouble loading this lab try using a different browser. We found Internet Explorer or Safari to work well.

Challenge 1: Alien Juice Bar

- Click on **Challenge 1** and then **Start**.
- Add cabbage juice indicator to the drinks and write down what color they turn.
- Place the drinks on the shelf that you think they belong on. Hit "check me" when you think you have it right. Fill in the data table when you get the bottles all right.

Drink	Acid, Base or Neutral?	Color with Indicator

Click on "Test More," add the indicator to each drink and correctly sort them on the shelves. List the drinks correctly in the data table below.

Acid	Neutral	Base

Challenge 2: The Flying Cabbage Juice Bar

- Go to the **Main Menu** and choose **Challenge 2** and then **Start**.
- Your customer will ask for an acidic, neutral or basic drink. Pour the cabbage juice indicator into the bottles of "juice" to figure out if they are acidic, basic or neutral.
- When you find what your customer wants, pour it into their cup and they will drink it. Be careful! If you give your customer the wrong drink, they will get sick. If they get the wrong drink a second time, they die! Fill in the data table below as you go.

	Drink Wanted	Type of "Juice" you gave.	Yummy or Sick?	Yummy or Dead?
Alien 1				
Alien 2				
Alien 3				
Alien 4				
Alien 5				
Alien 6				

Challenge 3: Alien Juice Bar

- Go to the **Main Menu** and choose **Challenge 3** and then **Start**.
- Put cabbage juice indicator into each of the cups. Add the different juices to make the cups acidic, basic or neutral. Fill in the data tables as you go.
- When all of your cups are the correct pH level, the game will automatically move on.

Acidic

Starting pH #						
Starting Color						
Ending pH #						
Ending Color						

Basic

Starting pH #						
Starting Color						
Ending pH #						
Ending Color						

Neutral:

Hint: This is a tough one! Different juices will raise or lower the pH in different amounts. Figure out the right combinations and you will be able to get your pH to the correct levels.

Starting pH #						
Starting Color						
Ending pH #						
Ending Color						

Questions:

1. What is pH?

2. Put a square around the pH number/s that are neutral, a circle around the numbers that are acidic and a line under the numbers that are basic.

   0 1 2 3 4 5 6 7 8 9 10 11 12 13 14

3. How would you define an indicator? Does an indicator just have to be for acids and bases?

4. When you are making an acid a base or a base an acid, this is a chemical reaction. It is called a neutralization reaction. What signs of a chemical change do you see as an acid or a base is neutralized?

Name:

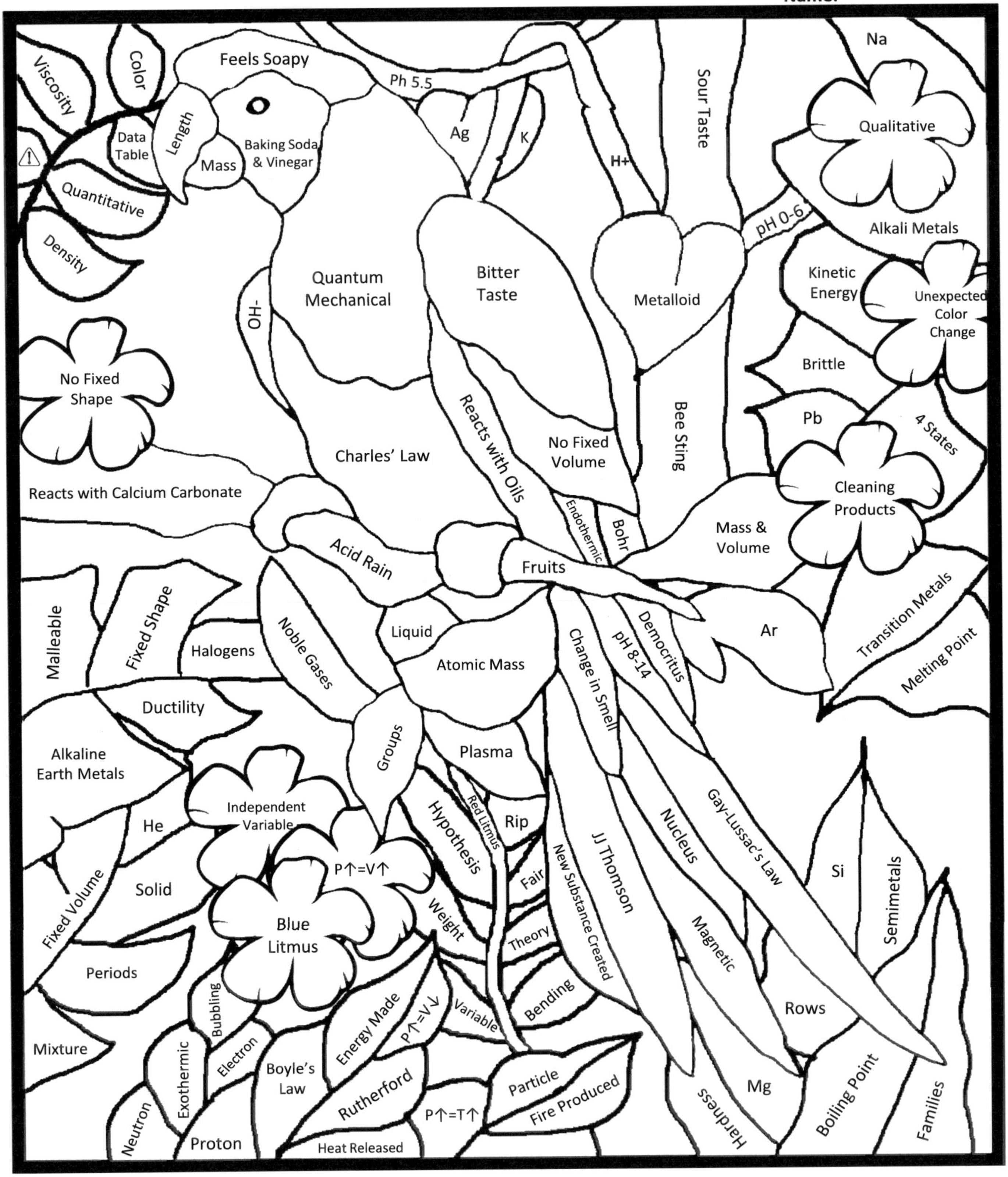

Color Chemistry Review

To find the color of each space, match it with the topic it best belongs to.

Bases-Dark Blue
Matter and its States-Light Green
Chemical Change-Light Blue
Gases-Purple
Atoms-Red
Acids-Brown
Physical Property-Yellow
Periodic Table-Green
Scientific Method and Safety-Orange

Invisible Ink Activity

An Acid Base Activity

Materials:

- Baking Soda
- Paper
- Water
- Paintbrush or Cotton Swab
- Frozen Grape Juice Concentrate

Caution: The grape juice concentrate can stain so make sure that you do not get it on clothes. This makes a non-toxic invisible ink using baking soda (sodium bicarbonate).

What to Do:

1. Mix 1/4 cup baking soda with 1/4 cup of water, in a cup or bowl.
2. Use a paintbrush or cotton swab to paint your secret message on a piece of paper.
3. Let this completely dry.
4. Use your paintbrush to paint grape juice concentrate across the paper. This will reveal your secret message!

What Happened:

Grape juice is an acid and it reacts with the baking soda which is a base. This reaction produces a different color so that your message is revealed.

Made in United States
Troutdale, OR
11/11/2026

57125326R00043